POWER TRANSACTION MANAGEMENT IN COMPETITIVE ELECTRICITY

by

MD SARWAR

ACKNOWLEDGEMENT

I would like to express my deep sense of gratitude and thanks to my supervisor Prof. Anwar Shahzad Siddiqui for continuous encouragement, invaluable guidance and wise counsel which helped me to conceptualize and manage research plan for the thesis work. Without his able guidance, it would have been impossible to complete the thesis in this manner. I consider myself extremely fortunate for being able to work and learn under him. My supervisor with his sharp inclusive intellect and deep insight of the subject has guided me on smooth and steady course. I express my gratitude to him.

I wish to express my gratefulness to Prof. Majid Jamil, Head, Department of Electrical Engineering, Jamia Millia Islamia, New Delhi for providing all necessary research facilities in the department.

I express my sincere gratitude to my parents, wife, sisters and other family members for their constant encouragement and support in completing this work. My endeavour for research work was continuously blessed by my parents Mr. Md Anwar Alam and Mrs. Sazda Khatoon.

I am wholly indebted to Almighty Allah who is omnipotent and superpower of the universe.

ABSTRACT

Power system is often operated upon its rated capacity that sometimes it results in loading of the transmission lines exceeding its thermal limits. This results in congestion of transmission lines. Congestion of transmission lines not only threatens the security of power system but it also affects its economic operation. Therefore, congestion management remains the central issue to the management of power transaction in competitive environment of electricity market. Market participants do not bother about the reliability and security of the system. They participate with the only aim of maximizing their profit. Without robust congestion management strategy, their actions can put the transmission system operation in stake. Therefore, interaction of congestion management with energy market economics should be carefully accomplished such that market inefficiency cannot take away the benefits promised by deregulation to the society. The management of congestion in transmission lines is somewhat complex in restructured power system as compared to that in Vertically Integrated Utility. The proposed work is focused on the different techniques adopted to manage congestion efficiently in competitive electricity market such that power system economics remain intact. Three different means of managing congestion have been proposed which are by using, i) generation rescheduling ii) distributed generation and iii) FACTS devices. Also, an efficient optimization technique based on Particle Swarm Optimization technique has been proposed to solve the non-linear congestion management problem. This thesis is also aimed to shed some light on various developments in electricity market structure.

Congestion management using generation rescheduling has been done by rescheduling the real power output of generators which participated in congestion alleviation. Sensitivity of the generators to power flow on congested line has been calculated to select the generators to participate in management of congestion. Based on their generator sensitivity values, the participating generators reschedule their real power output. An effort has also been made to take the power system economics into account. To get the minimum generation rescheduling cost, the bids submitted by generators have been considered along with generators sensitivity in order to calculate the amount of real power rescheduling. The method is implemented on IEEE 14-bus system and IEEE 30-bus system. For IEEE 14-bus system, the rescheduling cost comes out to be $286.1/hr when all the generators participated in congestion management and $231.5/hr when one of its generators having low value of generator sensitivity does not participate. For IEEE 30-bus system the rescheduling cost is $1472.3/hr. The effect of bids submitted by generators for output power rescheduling has also been analyzed. The increase of bid of generator from $19/MW^2 to $22/MW^2 reduces the real power output rescheduling of corresponding generator to 11.9 MW and 11.1 MW from 12.8 MW and 12.4 MW for IEEE 14-bus system and IEEE 30-bus system respectively. Accordingly, the other generators reschedule their power output to keep the generation rescheduling cost minimum.

A zonal congestion management scheme employing distributed generation (DG) has been presented in the thesis. The identification of different congestion zones is done on the basis of LMP difference across a line connected between two buses. The buses connecting lines of high and non-uniform LMP difference across them is grouped in most sensitive congestion zone while other buses connecting a line having low and uniform LMP difference are grouped in less sensitive zones of congestion. The DG is placed in the most sensitive congestion zone. Further, to find its optimal placement in most sensitive congestion zone, LMP difference is again utilized. The optimal placement of DG based on proposed method gives $3221.0/hr and $41638.6/hr as system generation cost for IEEE 14-bus system and IEEE 57-bus system respectively. The DG is also placed to other potential location within the most sensitive congestion zone and in other zones which gives higher system generation cost as compared to that of previous location. The optimal placement of DG also reduces the system congestion cost to $198.1/hr for IEEE 14-bus system as compared to $198.1/hr when the system is congested. Similarly, it reduces to $3584.4/hr from $4610.1/hr for IEEE 57-bus system.

A new congestion management method employing a series FACTS device has been proposed. The series FACTS device is placed optimally based on line flow sensitivity index. The optimal placement of the FACTS device is found to achieve its minimum installation cost as well as minimum generation cost of the system. For this, a static model of TCSC is formulated and implemented in the problem. The line flow sensitivity factor with respect to control parameter of TCSC is calculated for each line based on which TCSC is optimally placed in line having most negative line flow sensitivity factor. The proposed method has been tested on IEEE 14-bus system. It has been observed from test results that the placement of TCSC in most sensitive line determined from sensitivity analysis gives minimum installation cost of TCSC as $227.1/KVAR as compared to other potential locations of TCSC placement. Also the total generation cost found is less as compared to other potential locations for TCSC placement.

A reliable and efficient optimization technique to solve the non-linear congestion management problem has been implemented. PSO has been utilized in above reported work to find the amount of generation rescheduling and TCSC control parameter value. However, the selection of parameters in PSO plays an important role in its performance. Therefore, a more efficient optimization technique based on PSO is proposed to solve the congestion management problem. The new technique is termed as PSO with improved time varying acceleration coefficients (PSO-ITVAC). It is implemented in congestion management technique using generation rescheduling in order to analyze its performance and it has been observed that the problem converges to a more optimal solution with less number of iterations as compared to PSO with time varying acceleration coefficients (PSO-TVAC).

TABLE OF CONTENTS

Declaration	ii
Certificate	iii
Acknowledgement	iv
Abstract	v
Table of Contents	ix
List of Figures	xiii
List of Tables	xv
List of Abbreviations	xvii
List of Symbols	xx

CHAPTER-1 INTRODUCTION

1.1	Introduction		1
1.2	Deregulation of power systems		2
	1.2.1	Motivations for deregulation	2
	1.2.2	Milestones of deregulation: An historical background	4
	1.2.3	Deregulation scenario in India	6
	1.2.4	Challenges in deregulation	8
1.3	Congestion management		9
	1.3.1	Congestion management techniques	10
1.4	Organization of thesis		13

CHAPTER-2 LITERATURE SURVEY

2.1	Introduction	16
2.2	Literature review on competitive electricity market	17
2.3	Literature review on congestion management using generation rescheduling	19
2.4	Literature review on congestion management using FACTS devices	22
2.5	Literature review on congestion management using distributed generations	25

	2.6	Conclusion		27

CHAPTER-3	**COMPETITIVE ELECTRICITY MARKET: OVERVIEW AND DESIGN**			
	3.1	Introduction		29
	3.2	Market Structure and operation		30
		3.2.1	Objectives of market operation	30
		3.2.2	Electricity market models	30
		3.2.3	Competitive market design	32
	3.3	Power transaction management issues		34
	3.4	Conclusion		38

CHAPTER-4	**CONGESTION MANAGEMENT USING GENERATION RESCHEDULING**			
	4.1	Introduction		40
	4.2	Congestion Management using Generation Rescheduling		41
		4.2.1	Generator sensitivity	42
		4.2.2	Problem formulation for congestion management	43
		4.2.3	Selection of participating generators for congestion management	44
		4.2.4	Particle swarm optimization	46
		4.2.5	Congestion management algorithm using PSO	47
	4.3	Results and discussions		49
		4.3.1	Results for IEEE 14-bus system	49
		4.3.2	Results for IEEE 30-bus system	58
	4.4	Conclusion		66

CHAPTER-5	**CONGESTION MANAGEMENT USING DISTRIBUTED GENERATION**			
	5.1	Introduction		67
	5.2	Zonal based congestion management using distributed generation		67

		5.2.1	Locational marginal price	68
		5.2.2	Problem formulation	69
		5.2.3	Congestion zones identification	71
			5.2.3.1 Average LMP method	71
			5.2.3.2 LMP difference method	71
		5.2.4	Optimal allocation of distributed generation	72
			5.2.4.1 Highest LMP method	73
			5.2.4.2 LMP difference method	73
	5.3	Results and discussions		74
		5.3.1	Results for IEEE 14-bus system	75
		5.3.2	Results for IEEE 57-bus system	79
	5.4	Conclusion		84

CHAPTER-6	CONGESTION MANAGEMENT USING FACTS DEVICE		
	6.1	Introduction	86
	6.2	Basic concepts of FACTS	87
		6.2.1 An overview	87
		6.2.2 Classification of FACTS devices	89
	6.3	Thyristor Controlled Series Capacitor(TCSC): A series FACTS Device	90
		6.3.1 Modelling and implementation of TCSC	92
	6.4	Problem formulation for congestion management using TCSC	94
	6.5	Optimal placement of TCSC	96
	6.6	Results and discussions	99
	6.7	Conclusion	109

CHAPTER-7	PARTICLE SWARM OPTIMIZATION WITH IMPROVED TIME VARYING ACCELERATION COEFFICIENTS FOR CONGESTION MANAGEMENT	
	7.1 Introduction	110
	7.2 Particle swarm optimization with improved time varying	112

                acceleration coefficients (PSO-ITVAC)

7.3	Congestion management algorithm using PSO-ITVAC		116
7.4	Results and discussions		116
	7.4.1	Results for IEEE 30-bus system	117
	7.4.2	Results for IEEE 118-bus system	119
	7.4.3	Results for 33-bus Indian network	121
	7.4.4	Performance characteristics of PSO-ITVAC	123
7.5	Conclusion		127

CHAPTER-8   CONCLUSION AND FUTURE SCOPE

8.1	Introduction	128
8.2	Contributions of the research	129
8.3	Scope for future research	131

LIST OF FIGURES

Figure 1.1	Indian power market before deregulation	6
Figure 3.1	Competitive electricity market structure	34
Figure 4.1	Flowchart for congestion management based on PSO	49
Figure 4.2	Power flow for IEEE 14-bus system before generation rescheduling	52
Figure 4.3	GS values for IEEE 14-bus system	53
Figure 4.4	Power flow for IEEE 14-bus system after generation rescheduling (case I)	57
Figure 4.5	Power flow for IEEE 14-bus system after generation rescheduling (case II)	57
Figure 4.6	Effect of change of bid on generation rescheduling cost for IEEE 14-bus system	60
Figure 4.7	Power flow for IEEE 30-bus system before generation rescheduling	61
Figure 4.8	GS values for IEEE 30-bus system	62
Figure 4.9	Line flow for IEEE 30-bus system after generation rescheduling	65
Figure 4.10	Effect of change of bid on generation rescheduling cost for IEEE 30-bus system	66
Figure 5.1	Congestion zones identification based on LMP difference for IEEE 14-bus system	78
Figure 5.2	Generation cost for IEEE 14-bus system	80
Figure 5.3	Congestion zones identification based on LMP difference for IEEE 57-bus system	83
Figure 5.4	Generation cost for IEEE 57-bus system	85
Figure 6.1	Power flow between two systems	88
Figure 6.2	Functions of different FACTS devices	89
Figure 6.3	Thyristor Controlled Series Capacitor (TCSC)	91
Figure 6.4	Impedance vs firing angle characteristic of TCSC	92
Figure 6.5	Transmission line model	93
Figure 6.6	Transmission line model with TCSC	94

Figure 6.7	Power injection model	95
Figure 6.8	Flow chart for optimal setting of TCSC	99
Figure 6.9	Line Flow Sensitivity Factor for IEEE 14-bus system	102
Figure 6.10	Power flow for IEEE 14-bus system with TCSC in line-7	106
Figure 6.11	Power flow for IEEE 14-bus system with TCSC in line-1	107
Figure 6.12	Power flow for IEEE 14-bus system with TCSC in line-4	107
Figure 6.13	Installation cost of TCSC	108
Figure 6.14	Total generation cost for IEEE 14-bus system	109
Figure 7.1	GS values for IEEE 30-bus system	119
Figure 7.2	GS values for IEEE 118-bus system	121
Figure 7.3	GS values for 33-bus Indian Network	123
Figure 7.4	Rescheduling cost and active power rescheduling for different systems	124
Figure 7.5	Convergence characteristics of PSO-ITVAC and PSO-TVAC for IEEE 30-bus system	127
Figure 7.6	Convergence characteristics of PSO-ITVAC and PSO-TVAC for IEEE 118-bus system	127
Figure 7.7	Convergence characteristics of PSO-ITVAC and PSO-TVAC for 33-bus Indian network	128

LIST OF TABLES

Table 4.1	PSO parameters	50
Table 4.2	Power flow results for IEEE 14-bus system	51
Table 4.3	Generator sensitivity values of IEEE 14-bus system for congested line between bus-2 and bus-3	52
Table 4.4	Rescheduling bids submitted by different generators of IEEE 14-bus system	54
Table 4.5	Rescheduling results for IEEE 14-bus system	54
Table 4.6	Power flow results for IEEE 14-bus system for case I	55
Table 4.7	Power flow results for IEEE 14-bus system for case II	56
Table 4.8	Effect of change of bids on generation rescheduling for IEEE 14-bus system	58
Table 4.9	Power flow results for IEEE 30-bus system	59
Table 4.10	Generator sensitivity values of IEEE 30-bus system for congested line connected between bus-1 and bus-2	61
Table 4.11	Rescheduling bids submitted by different generators of IEEE 30-bus system	63
Table 4.12	Rescheduling results for IEEE 30-bus system	63
Table 4.13	Power flow results for IEEE 30-bus system after generation rescheduling	63
Table 4.14	Effect of change of bids on generation rescheduling for IEEE 30-bus system	66
Table 5.1	LMP difference across lines for IEEE 14-bus system	76
Table 5.2	Congestion zones identification based on LMP difference for IEEE 14-bus	77
Table 5.3	Results for IEEE 14-bus system	78
Table 5.4	Generation cost for IEEE 14-bus system in different zones	79
Table 5.5	LMP difference across lines for IEEE 57-bus system	81
Table 5.6	Congestion zones identification based on LMP difference for IEEE 57-bus	82

Table 5.7	Results for IEEE 57-bus system	84
Table 5.8	Generation cost for IEEE 57-bus system in different zones	84
Table 6.1	Classification of FACTS devices	90
Table 6.2	Power flow for IEEE 14-bus system without TCSC	100
Table 6.3	Line flow sensitivity factor of IEEE 14-bus system for congested line-3	101
Table 6.4	Power Flow for IEEE 14-bus system with TCSC in line-7	103
Table 6.5	Power flow for IEEE 14-bus system with TCSC in line-1	104
Table 6.6	Power flow for IEEE 14-bus system with TCSC in line-4	105
Table 6.7	Total generation cost for IEEE 14-bus system	108
Table 6.8	Total generation cost for IEEE 14-bus system	109
Table 7.1	PSO-ITVAC parameters	118
Table 7.2	Generator sensitivity values of IEEE 30-bus system for congested line connected between bus-1 and bus-2	119
Table 7.3	Results comparison for IEEE 30-bus system	119
Table 7.4	Generator sensitivity values of IEEE 118-bus system for congested line connected between bus-89 and bus-90	120
Table 7.5	Results comparison for IEEE 118-bus system	121
Table 7.6	Generator sensitivity values of 33-bus Indian network for congested line connected between bus-8 and bus-23	122
Table 7.7	Results comparison for 33-bus Indian network	123
Table 7.8	Statistical results for IEEE 30-bus system	125
Table 7.9	Statistical results for IEEE 118-bus system	125
Table 7.10	Statistical results for 33-bus Indian network	126

LIST OF ABBREVIATIONS

ACO	Ant Colony Optimization
ANN	Artificial Neural Network
BSO	Bacteria Swarming Optimization
CalPx	California Power Exchange
DE	Differential Evolution
DG	Distributed Generation
DISCO	Distribution Company
EP	Evolutionary Programming
FACTS	Flexible Ac Transmission System
FL	Fuzzy Logic
FLV	Line Flow Limit Violation Factor
FTR	Financial Transmission Rights
GA	Genetic Algorithm
GENCO	Generation Company
GS	Generator Sensitivity
GSA	Gravitational Search Algorithm
HS	Harmony Search
IEEE	Institute of Electrical and Electronics Engineers
IPFC	Interline Power Flow Controller
IPP	Independent Power Producers
ISO	Independent System Operator
LMP	Locational Marginal Price
LP	Linear Programming

LSF	Line Flow Sensitivity Factor
MILP	Mixed Integer Linear Programming
MINLP	Mixed Integer Non-Linear programming
MVA	Mega Volt Ampere
MVAR	Mega Volt Amperes Reactive
MW	Mega Watt
NGC	National Grid Company
NLP	Non-Linear Programming
OPF	Optimal Power Flow
PSO	Particle Swarm Optimization
PSO-ITVAC	Particle Swarm Optimization with Improved Time Varying Acceleration Coefficients
PSO-TVAC	Particle Swarm Optimization with Time Varying Acceleration Coefficients
PTC	Power Trading Corporation of India
QP	Quadratic Programming
SA	Simulated Annealing
SEB	State Electricity Board
SSSC	Static Synchronous Series Compensator
STATCOM	Static Synchronous Compensator
SVC	Static Var Compensation
TCPAS	Thyristor Controlled Phase Angle Shifter
TCPS	Thyristor Controlled Phase Shifter
TCSC	Thyristor Controlled Series Compensation
TCVR	Thyristor Controlled Voltage Regulator

TRANSCO	Transmission Company
TS	Tabu Search
UPFC	Unified Power Flow Controller
VLV	Voltage Violation Factor
VSC	Voltage Source Converter

LIST OF SYMBOLS

α	Firing angle
λ	Lagrangian multipliers vectors associated with equality constraints
λ_1	Penalty coefficients in the range of 10^5 to 10^8
λ_2	Penalty coefficients in the range of 10^5 to 10^8
μ	Lagrangian multipliers vectors associated with inequality constraints
θ_k	Phase angle at bus-k
θ_p	Phase angle at bus-p
θ_q	Phase angle at bus-q
θ_s	Phase angle at bus-s
δ_i	Volatage angle at bus-i
δ_j	Volatage angle at bus-j
δ_{ij}	Phase angle difference between two bus voltages
ΔLMP_{kl}	LMP difference across line-kl,
ΔLMP_L	LMP difference across line-l,
ΔP_{gi}	Active power adjustment of generator at bus-i
ΔP_{gi}^{max}	Maximum limit of active power adjustments of generator at bus- i
ΔP_{gi}^{min}	Minimum limit of active power adjustments of generator at bus- i
B_{pq}	Susceptance of line connected between bus-p and bus-q
B_{sk}	Susceptance of line connected between bus-s and bus-k
c_1	Acceleration coefficients

c_2	Acceleration coefficients
CC_m	Congestion component of LMP at bus m
C_i	Cost of generation for generator i
C_t	Unit cost of TCSC
C_{TCSC}	Installation Cost of TCSC
F_L	Real power flow on transmission line-l
F_L^{max}	Real power flow limit of line-l
GS_i^{pq}	Generator sensitivity (GS) of generator at bus-i
G_{pq}	Conductance of line connected between bus-p and bus-q
G_{sk}	Conductance of line connected between bus-s and bus-k
LC_m	Loss component of LMP at bus-m
LMP_k	LMP at bus-k
LMP_l	LMP at bus-l
LMP_m	LMP at bus-m
LMP_n	LMP at bus-n
LSF_c^k	Line flow sensitivity factor with respect to the parameters of TCSC placed at line-k
MEC_m	Marginal energy component of LMP at bus-m
n_b	Number of buses
n_g	Total Number of generators
n_L	Total number of lines in the system
n_L	Total number of lines
P_{D_k}	Real Power demand at bus- k
P_{gi}	Active power generation of generator at bus-i

P_{G_k}	Real Power Generation at bus-k
P_{Gk}^{max}	Maximum real power output limits of kth generator
P_{Gk}^{min}	Minimum real power output limits of kth generator
P_i	Active Power injected at bus-i
P_L	Real power flow in line-k to which TCSC is connected
P_{LT}	Real power flow in line connected between bus-i and bus-j
P_{ij}	Active power flow from bus-i to bus-j
P_{pq}	Active power flow on congested line-l connected between bus-p and bus-q
Q_1	Reactive power flow in the line before installation of TCSC.
Q_2	Reactive power flow in the line after installation of TCSC.
Q_{Gk}^{max}	Maximum reactive power output limits of kth generator
Q_{Gk}^{min}	Minimum reactive power output limits of kth generator
Q_{ij}	Reactive power flow from bus-i to bus-j
r_1	Random values between 0 and 1
r_2	Random values between 0 and 1
r_{ij}	Resistance of a line connected between bus-1 and bus-j
RC_i	Rescheduling cost of generator-i
V_1	Voltages at bus-1
V_2	Voltages at bus-2
V_b	Voltages at bus-b
V_i	Voltages at bus-i
V_j	Voltages at bus-j
V_k	Voltages at bus-k

V_m^{max}	Maximum voltage limits at bus-m.
V_m^{min}	Minimum voltage limits at bus-m.
V_p	Voltages at bus-p
V_q	Voltages at bus-q
V_{ref}	Reference voltages at a bus
V_{ref}^{max}	Maximum reference voltage at a bus
V_{ref}^{min}	Minimum reference voltage at a bus
w	Inertia weight
w_{min}	Minimum value of inertia weight
w_{max}	Maximum value of inertia weight
X_C	Capacitive reactance of a line
X_{ck}	Control parameter of TCSC
X_L	Inductive reactance of a line
x_{ij}	Reactance of a line connected between bus-1 and bus-j

CHAPTER – 1

INTRODUCTION

1.1 Introduction

With the advancement of technology as well as population throughout the world, the demand of electricity has increased in last few decades. Globally, the per-capita consumption of electricity, a parameter to gauge the development and economic growth of a country, has increased sharply. The growth in electrical energy consumption has kept the required demand unfulfilled. Although a large amount of electrical energy has been added yearly to the installed capacity, the gap in demand and supply is increasing every year. The growth in demand of electrical energy has been so phenomenal that it exceeds the supply by a handsome margin. The growth and development of generation, transmission and distribution sectors have not kept pace with the demand of electricity. Even the development amongst generation, transmission and distribution has been skewed. The power networks have also grown more complex. All these factors have led to reduced security, stability and reliability of power system. These factors are further aggravated by deregulation of vertically integrated power sector which may overload the transmission corridors due to power wheeling, unplanned exchange etc. This may put the security and stability of the system in danger. The issues such as environmental, economic and most importantly right of way have made it essential to use the existing infrastructure optimally so that the electricity market, created by deregulation of power system, operate efficiently and the benefits promised by deregulation to the society remains intact. To achieve this, a number of developments have taken place since deregulation of electric power system.

1.2 Deregulation of power systems

Over the years, electricity industry is operated as a single utility, known as vertically integrated utility, which has the control over all activities of power system (generation, transmission and distribution) within its operational domain. This single utility is accountable to provide electricity in its region. Their prime and only objective is to minimize the total system cost while all associated system constraints being satisfied. Such operational limitation as well other technological and commercial issues have caused enormous changes in electricity industry during last two decades. It evolves into a distributed and competitive industry in which price of electricity is driven by market forces and the cost of electricity is reduced through increased competition [1].

Deregulation of power system has decomposed its three components: generation, transmission and distribution into three separate and independent business entities. It introduces competition in the electricity industry through participation of private players which leads to service improvement, competitive electricity price, and the availability of capital for up-gradation and expansion of system infrastructure. The end consumers get benefited with low price of electricity delivered [1].

1.2.1 Motivations for deregulation

The restructuring of the electricity industry around the world has taken place at a rapid pace. The key features which contribute in the restructuring process were participation of independent power producers and distributors to decrease the electricity price, improve system efficiency, reduce excess manpower and allow greater alternatives to consumers. So, deregulation has evolved as a panacea for mismanaged, financially crunched and inefficient vertically integrated power system [1].

The major reasons for deregulation are following-

(i) High electricity cost- Due to monopoly of state controlled powers utilities, the price of electricity was high. This high price of electricity was expected to decrease due to entry of private players with deregulation.

(ii) Managerial Inefficiency- The interference of the government and delay in taking decisions by the management had made the system inefficient. This managerial inefficiency was eliminated with introduction of competition due to deregulation.

(iii) Customer choice – Monopoly in electricity market made the consumer to get electricity from only available service provider. But, deregulation has provided the consumers with an option of buying electricity from supplier of their choice at a competitive rate.

(iv) Future development – Almost all utilities were facing the problem of financial crisis due to losses. They lacked in necessary financial resources to carry out expansion and development of existing infrastructure.

(v) Utilization of resources – In vertically integrated system, due to lack in competition, the available resources are not fully utilized. But the deregulation came up with a solution for utilities to use the available resources in a better and optimum way.

(vi) Financial Help – The financial help was available to utilities from the financial institutions such as World Bank, Asian Development Bank etc. to take the necessary restructuring reforms in electrical industry and improve its condition after deregulation.

(vii) Customer Service – The inefficient operation of vertically integrated utility did not focus on the customer service. A better customer service is achieved through deregulation.

(viii) Deregulation of other sector- The positive experience gained from the deregulation of other sectors such as telecom, airline, TV and broadcast, manufacturing etc. has given a way for the adoption of deregulation in electrical industry.

1.2.2 Milestones of deregulation: An historical background

Pre-deregulation, all power utilities were owned and managed by government with monopoly in operation. This deteriorated the condition of utility due to various reasons such as mounting losses, corruption, system lethargy, unavailability of financial resources, no thrust for modernisation and expansion etc. Also, the utility was unable to meet the increasing demand. The monopoly in its operation also leads to poor service to customers and underutilisation of available resources. All these had compelled the government to give a rethink on the existing policies and resulted in the process of deregulating the electricity sector which would introduce the private players in it. This was expected to improve operational efficiency and service quality. As a result, power utilities were unbundled into different independent business entities which are Generation Company (GENCO), Transmission Company (TRANSCO) and Distribution Company (DISCO).

Chile was first to initiate this game changing transformation which privatised its electric utility in 1982. The transmission from regulated to deregulated structure was smooth due to well planned tariff structure and stable govt. This success provided momentum for privatisation of other utilities. So, privatisation became a tool for deregulation.

U.K followed the footsteps of Chile and transformed its electric industry in 1990 with the enforcement of Electricity Act. The monopolistic structure was unbundled to create privately owned generation companies and privately owned regional electricity companies to distribute power while the transmission system was owned and operated by National Grid Company (NGC).

In 1992, Argentina modified the Chilean model of deregulation and prevented the market dominance by restricting the concentration to few large players. This had resulted in more efficient operation of electricity market. The outcome of deregulation process in Latin America was so encouraging that it forced Peru, Columbia and Bolivia to deregulate their market in 1993 and followed by Australia in 1994 which set up wholesale spot market.

New Zealand restructured its electricity sector in 1996 by setting up independent system operator and established a wholesale electricity market. It utilized the Financial Transmission Rights (FTR) as a tool to hedge congestion.

In 1998, U.S.A. had transformed the wholesale generation into a competitive market with open access to transmission [2]. California was first to go for deregulation followed by Massachusetts and New York. The Californian model consisted of open access in transmission with competitive forward market [3]. The auctions were held through a centralized independent entity called as California Power Exchange (CalPX). But the deregulation process was not a success due to concentration of market powers into hands of few sellers during congestion. However, the deregulation model adopted by Pennsylvania – Jersy – Maryland (PJM) was a success [4].

In Asian region, Japan was first to initiate the deregulation process in 1995 with the establishment of wholesale market and Independent Power Producers (IPP). It later established a power exchange in 2003. Nigeria initiated the deregulation process in 1999 and was first country in African continent to adopt it.

1.2.3 Deregulation scenario in India

In India, the consumers had been long served by vertically integrated State Electricity Boards (SEBs) where in a single entity is responsible for operation and control of power system as well as setting of electricity tariff as shown in Figure 1.1. But due to inefficiency of the vertically integrated model and lack of infrastructure, some of the SEBs were forced by foreign funding agencies to restructure their power sector, resulting in separation of generation, transmission and distribution as an independent entity.

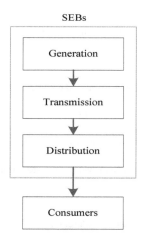

Figure 1.1 Indian power market before deregulation

The deregulation process in India started way back in 1991 with participation of private players in generation [5]–[9]. The existing electricity act was revised to allow participation of private sector in operation and maintenance of power generation companies. Further, to achieve the smooth transition from regulated to deregulated structure, a regulatory body was established through the enactment of Electricity Regulatory Commission Act, 1998. However, the milestones in deregulation of its electricity sector were achieved by the enactment of the Electricity Act, 2003. The

Electricity Act 2003 laid down provisions to transform a regulated electricity market towards a competitive electricity market and hence opened the power sector. The act identified the electricity trade as a separate activity which paved the way for a paradigm shift in generation, transmission and distribution activities. This helps in promoting economy, efficiency and investments in power sector. The act provides an open and non-discriminatory access of centre as well as state owned transmission network, thereby promoting the active participation of private as well as state owned generators and industrial consumers [9].

Orissa was the first state in India to restructure its state electricity board in 1996 in order to come out of reeling condition of its electricity sector. This reform was funded by the World Bank. However, deregulation in Orissa came as a failure [9]. But the other states took lessons from it and came up with the solution of reasons of failure and subsequently deregulated their State Electricity Boards. Haryana restructured its electricity board in 1998 while Andhra Pradesh and Karnataka had adopted it in 1999 [10]. Delhi and Rajasthan were next to adopt deregulation in 2002. With the enactment of Electricity Act 2003, other states such as Assam, Maharashtra, West Bengal and Gujarat adopted deregulation in between 2004 to 2007. Later, it was followed by Chhattisgarh, Jharkhand, Punjab, Tamil Nadu and Bihar after 2008 [11].

The state electricity regulatory commissions, controlled by Central Electricity Regulatory Commission set up in1991, were established in each state to solve the issue of business licensing and tariff determination. It also sets up trade regulations to allow open access of transmission systems.

Another milestone in deregulation of Indian electricity sector was achieved with the establishment of power trading activities through Power Trading Corporation of India (PTC) in June, 2002. The work of PTC was to maintain the supply-demand balance by purchasing power from surplus region and selling it in deficit area. The aim is to act as a service provider in power transaction activities without owning any generation or transmission network. It participates in future trading by entering into agreement with multiple buyers and sellers. In fact, for power trading with other countries such as Bhutan, Nepal and Bangladesh, PTC is designated as a nodal agency.

1.2.4 Challenges in deregulation

The benefits promised by deregulation of electricity market to the society make it a viable choice to adopt. However the deregulated electricity market presents a more complex and dynamic scenario than vertically integrated utilities. This is due to the existence of multiple contracts amongst market participants, variety of market model, management of ancillary services, etc. These issues of deregulated electricity market may lead to its failure (e.g. California in U.S.A and Orissa in India). However, experience gained from deregulated market helped to solve some of the issues to some extent. These issues are both technical as well as financial in nature. Some of the key issues are given as-

- Congestion management.
- Price volatility.
- Transmission Pricing.
- Ancillary service management etc.
- Available transfer capacity calculation.
- Market power.

These are some of the issues of deregulated electricity market that need to be handled in appropriate way so that they cannot take away the benefits promised by deregulation to the society.

1.3 Congestion management

Congestion is defined as the inability of transmission system to transfer power through a certain path due to violation of transfer limits such as stability limit, voltage limit and thermal loading. It may also occur if economic restrictions, such as contract enforcement, priority feed in etc., prevent further transfer of power through a certain transmission path. It is a condition arises due to the utilization of available transfer capacity of the transmission path and no more transmission can take place without the violation of either technical constraints or economic restrictions. The existence of such a condition for a longer duration may endanger the security and reliability of the system. It may also leads to line or power equipment outage, loss of revenue, diminished security and stability, out of merit dispatch, pay contractual penalty, poor power quality and higher price of electricity to consumers.

The main reasons which cause the system to become congested are given below-

(i) *Thermal limit-* It is an upper limit of current that a line can carry for a fixed duration. If the current limit is exceeded, the transmission line conductors suffer permanent damage due to excessive heat generated. This can be caused by both active as well as reactive power flow.

(ii) *Voltage limit-* The magnitude of voltage is maintained between allowable upper and lower limit. The amount of power transferred is affected by allowable lower limit of system voltage.

(iii) *Stability limit-* The stability refers to the ability of a system to remain in synchronism when subjected to disturbance. The duration of disturbance could be few milliseconds to several minutes. The disturbance causes the generators in the local region to oscillate with respect to one another affecting voltage, frequency and line loading. For a system to stabilize, the oscillations must damp out fast.

Apart from above reasons, there are other implications of congestion in the market operation. Congestion may force the system operator to re-dispatch in some cases which will affect the market economics. It will affect the market players with some market players come as a winners while some as losers. Also, some market players may manipulate the market for their advantage by exploiting the situation caused due to congestion.

Hence, management of congestion is one of the most important and challenging tasks of system operator so that the system remains secure and reliable while maintaining the market economics intact.

1.3.1 Congestion management techniques

Deregulation of electricity market has provided an open access of transmission system to all market participants. The power flow gets influenced by generation pattern, contingency conditions, extreme weather conditions, cyclical variations etc. which may leads to system congestion. To prevent the system from becoming unsecure, it is essential to establish certain congestion management system consisting of rules and procedure such that it maintains the electricity market economics.

Different congestion management schemes are adopted by electricity markets. The choice of congestion management scheme depends upon the network topologies and demographic factors as well as political ideologies. Each method of congestion management has the prime objective of maintaining system security. However, they have different economic impact on electricity market. Hence, any particular congestion management methodology cannot be said to be suitable to all market structure. The method of congestion management should be transparent, clear and robust.

The various schemes of congestion management can be broadly categorised as follows:

(i)   Market based methods.
(ii) Non-market based methods.
(iii) Technical methods.

(i) *Market based methods-* These methods can further be subdivided into following categories-

   (a) *Explicit auctioning-* This method involves the auction of available transmission capacity of tie lines to highest bidder. The disadvantage of this method is that it does not consider loop flows and makes trading activities complicated.

   (b) *Nodal pricing or optimal power flow based congestion management-*

   The Optimal Power Flow (OPF) solution gives the optimum power transfer while fulfilling all network constraints. The optimization gives a power price at each node which is known as nodal price. Nodal price is defined as the variation in price energy at each node with respect to demand at that node. OPF is a mathematical formulation which combines economic dispatch with power flow equations subject to various inequality and equality constraints. The objective function could be – minimization of load curtailment, minimization of cost of generation subject to voltage with given tolerance band, reactive power flow between a maximum and minimum value etc.

   (c) *Zonal pricing* – This method combines system buses into a zone having same Locational Marginal Price (LMP). The market is initially settled without constraints and if congestion is detected, supplementary bids for decrease or increase of generation are invited. The zonal price is decided by lowest supplementary bid for reduction in generation.

(d) *Market splitting or price area congestion management* – In the absence of any congestion, the network is scheduled without constraints. However, if congestion occurs, the market is split and separately settled. The system operator purchase power from low price generator having excess power and sells it to high demand area at high price. The difference in revenue earned by ISO is used for transmission infrastructure development. The main advantage offered by this method is that it encourages generating companies to invest in building new capacities in high demand zones so that competition could be introduced. However, this method cannot be applied to power networks having no well-defined boundaries.

(e) *Re-dispatch* – In case of occurrence of congestion, the system operator uses generation re-dispatch and purchases power at higher price to be sold at lower price to mitigate congestion.

(ii) *Non-market based methods* – These methods are based on allocation of scarce transmission capacity on some criteria.

(a) *First come, first served method* – The available transmission capacity is offered to the first party who requests it and the distribution go on till the entire capacity is exhausted.

(b) *Pro-rata basis method* – This method allocates a certain share of the available capacity to the parties. The share is ratio of total available capacity to sum of requests received. This method allows some transmission capacity offered to all interested parties. Generally, developing countries follow this model.

(c) *Curtailment of load* – This method uses proportionate reduction of load to reduce congestion. This leads to reduced price in the area.

(d) *Counter-trade* – The congestion can be relieved if power is traded in a direction opposite to the direction of congestion. This method could be termed as modified re-dispatch.

(iii) *Technical methods* – These methods minimize congestion by changing topology of the power network. These methods are as follows:

(a) *Use of transformer taps* – By changing taps, voltage can be increased to reduce congestion. However, this method gives small relief.

(b) *Outaging of congested line* – The line under congestion is switched off and load is curtailed. Though this method is seldom preferred.

(c) *Use of FACTS devices* – Flexible AC Transmission System or FACTS are power electronics devices which can regulate the power flow by altering either or a combination of transmission line parameters such as X (line reactance), V (voltage magnitude) and δ (voltage angle). These devices can also be used for voltage stability improvement, transient stability improvement, sub- synchronous resonance mitigation etc. These devices reduce congestion by increasing power flow in the existing lines.

1.4 Organisation of thesis

The present research work involves the management of congestion through generation rescheduling, FACTS device and distributed generations. The contribution of this thesis starts with mathematical modelling and programming in MATLAB to analyse the performance of above adopted schemes of congestion management on standard power system networks. The above cost free means as well as non-cost free means of congestion mitigation techniques have been evaluated in a pool type of electricity market. A new optimization algorithm based on particle swarm optimization technique has been introduced for optimal rescheduling of generators output in order to manage congestion efficiently and economically.

The work presented in this thesis is divided into eight chapters. The chapter wise summary of the thesis is given below-

Chapter – 1 presents introduction to power system deregulation, reasons behind deregulation, their historical evolution and status around worldwide and in India. It describes key issues in deregulation process. It also describes the various techniques available for congestion management.

Chapter – 2 presents literature review on market structure and congestion management schemes adopted in competitive environment of electricity market. It provides the review of literature on congestion management using generation rescheduling as well as by optimal placement of distributed generations and FACTS device. This lays down the objectives and motivation for the research work carried out.

Chapter – 3 investigates the various issues of power flow in a transmission path in competitive environment of electricity market. It deals with the various market structure of a competitive electricity market and market design adopted to solve the above investigated issues.

Chapter – 4 deals with congestion management using generation rescheduling. The amount of rescheduling depends upon the generators sensitivities as well as their bids submitted to the system operator. The problem is formulated as an Optimal Power Flow (OPF) problem in a pool market and is implemented in IEEE-14 bus system and

IEEE 30-bus system. Further the effect of variation of bids on amount of generation rescheduling for congestion management is also analysed.

Chapter – 5 focuses on use of distributed generations for managing congestion. It describes the zonal congestion management approach based on Locational Marginal Price (LMP) which is also used to find the optimal location for distributed generations placement in order to get low system generation cost. The method is implemented in IEEE-14 bus and IEEE-57 bus systems.

Chapter – 6 describes about the optimal placement of FACTS device for congestion mitigation. The simplified model of Thyristor Controlled Series Compensator (TCSC) is incorporated in load-flow algorithm. These devices are then implemented in IEEE-14 bus system.

Chapter – 7 focuses on a new optimization algorithm based on particle swarm optimization technique. The performance of the new optimization algorithm is evaluated for solving the congestion management problem using generation rescheduling. This method is implemented in IEEE 30-bus and IEEE 118-bus systems. It is also implemented in a 33-bus Indian network.

Chapter – 8 summarizes the conclusions and contributions of research work along with scope and suggestions for future research work.

CHAPTER – 2

LITERATURE SURVEY

2.1 Introduction

In the last three decades, almost all countries in the world have adopted the deregulation of their power sector [1]–[4]. With the initiation and success of power sector reforms by Chile in early eighties, every country in the world took a step forward to adopt the revolutionary reforms in their power sector. Since then deregulation of power sector has become a burning topic worldwide. In India, Orissa was the first state to unbundle its electricity board in 1996 followed by Haryana in 1998, Andhra Pradesh and Karnataka in 1999 and so on [5]–[11]. Due to the benefits promised by the restructuring of power sector, a lot of work has been done in this field of deregulation. Also, with the introduction of competition in electricity market due to deregulation, the power flow in transmission line has become difficult to manage. A large number of papers have appeared in the area of deregulation which addresses the issues of transmission line management.

The literature in the subject area of power management in competitive environment of power sector is so vast that in the present chapter only a modest attempt has been made to provide a review of significant work in the following areas:-

(i) Competitive electricity market.
(ii) Congestion management using generation rescheduling.
(iii) Congestion management using FACTS devices.
(iv) Congestion management using distributed generations.

The literature review on the above areas is discussed in the following sections.

2.2 Literature review on competitive electricity market

Introduction of deregulation in power sector has changed the existing structure of power system. It introduces an efficient, competitive and consumer oriented culture in the power sector. A new competitive market paradigm has evolved which is driven by market forces and strict environmental conditions [12]–[14]. Although a number of market structures [15]–[17] exist for trading of electrical energy, but all these market structures cannot be adopted arbitrarily by any country. The adoption of a specific electricity market structure depends upon suitability of a particular country considering several issues such as financial issues, topographical issues etc. Due to unbundling of power sector, different market structures are adopted to mitigate the issues related to open access of transmission lines.

Different electricity market structures have been discussed in [18]–[20]. With the deregulation of power sector, the adoption of appropriate electricity market structure becomes a major concern for the utilities. Fang et al in [18] have presented the possible emerging market structures that will exist in future due to deregulation. These market structures consist of pool transactions, bilateral/multilateral transactions and ancillary services transactions which are used for optimal transmission dispatch in an open access of transmission lines. In [19] similar kind of electricity market models have been discussed. The authors have also presented a hybrid electricity market model wherein PoolCo and bilateral transaction simultaneously exist. To solve the issue of transmission line congestion, Ashwani et al presented a review of the available market structures.

Post-deregulation the power system become unbundled and its different components are controlled independently by different entities. To provide a non-discriminatory access of transmission lines as well as secure reliable operation of power system, pool model of electricity market has been widely adopted. This market model has been discussed in [21]-[22]. A single side bid PoolCo market is proposed by Lu et al [21] to maximize the social welfare benefits. Hassan et al [22] compared a single buyer model of electricity market with single side bid PoolCo model in terms of generation revenue. In [23], the authors have adopted similar pool electricity market model for minimization of fuel and reactive power cost.

Besides the PoolCo model, another model known as bilateral transaction model of electricity market has also been adopted by different utilities. A market model having bilateral transaction is presented in [24]-[25].The authors have discussed the modeling of bilateral contracts by using a transaction matrix. The concept of the transaction matrix is to expand the conventional load flow variables. Since, with the introduction of competition in electricity market, a large number of transactions are expected to take place which would hamper the security of the system. These bilateral transactions need to be evaluated for their feasibility ahead of their scheduling time so that the system remains secure when these transactions take place simultaneously. Such a feasibility of simultaneous bilateral transactions have been accessed by a method proposed by Hamoud [26]. Kumar et al proposed such a feasible bilateral transactions using a secure bilateral transaction matrix which would keep intact the system security and reliability. Another aspect to schedule bilateral transactions has been considered by Galiana et al.[27]. The authors have proposed an electricity market model having bilateral transactions considering the impact of these transactions on system losses. Each bilateral transaction is allocated with a component of system losses in order to achieve higher economic efficiency. In [28], an approach to assess the different aspects of bilateral market has been presented. The bilateral trading patterns have been analyzed in the electricity markets considering the network physical constraints for secure operation of power systems. A bilateral contract negotiation using Nash bargaining theory for settlement between GENCO and DISCO has been analyzed in [29].

Although bilateral transaction model of electricity market is important, it attracts less attention due to complexity of the model in terms of trading patterns which reflects the self-interested behavior of the market participants to find potentially profitable transactions. Also, the unfeasible bilateral transactions also lead to insecure operation of power systems. Therefore, a market model with bilateral trading in presence of a pool trading attracts more attention. Such a model of electricity market is often called as hybrid electricity market [30]-[31]. Xiao et al. has discussed a hybrid electricity market model wherein market is cleared with uniform price in coordination with congestion free dispatch. Similar market structure has been modeled in [32] for congestion management. Kumar et al proposed a hybrid electricity market having secure bilateral transaction using AC distribution factors while in [33], the authors have modeled the hybrid electricity market with secure bilateral transaction using generators' up and down cost bids. A day- ahead hybrid electricity market is proposed in [34], with impedance, current and power load model. The proposed market structure is utilized for congestion mitigation scheme.

2.3 Literature review on congestion management using generation rescheduling

Congestion management of a transmission line is one of the major economic concerns of an electric power system. Transmission congestion occurs when no additional power can be transferred from injection to extraction point due to violation of its transfer limits. Thermal limits, stability limits and voltage limits are some of the main causes of congestion of a transmission line. Management of congestion means to simultaneously accommodate all the constraints of a transmission line. Before deregulation, congestion management is central to the dispatch and somewhat easier as all the components of power system are under the control of single entity. But after deregulation, the management of congestion becomes complex. The introduction of competition in electricity market makes it more complex to manage congestion as all the energy suppliers will try to transfer their generated electricity [35]. This would put the security of a power system in danger. Also there is considerable loss of economy due to congestion. Therefore, the management of congestion plays a vital role to keep the promises of a competitive electricity market intact. Congestion can be managed in a number of ways. However, in present day competitive electricity market, each utility manages the congestion in transmission network using its own rules and guidelines utilizing certain physical or financial mechanism. In [36]–[41] different congestion management schemes, suitable for competitive market structures have been discussed. Generation rescheduling, load curtailment, operation of on-load tap changers, use of FACTS devices etc. are generally used to manage congestion in competitive electricity market.

The congestion management using generation rescheduling [42]–[46] generally involves calculation of generator sensitivity factor based on Optimal Power Flow (OPF). A constant voltage profile is assumed in DC power flow method which is not always true. Moreover, a contingency such as line outage will cause the voltage profile to undergo major changes. These problems are somewhat eliminated in AC power flow approach and is more accurate than DC power flow method.

A congestion management approach using DC Optimal Power Flow (OPF) is proposed in [47]. The authors have presented a way for optimal dispatch of generators in order to deal with the transmission congestion cost in competitive electricity market with pool and bilateral transactions. A similar DC Optimal Power Flow approach employing the generator cost curve for congestion management has been utilized in [48]. The author has provided a tutorial review to calculate the congestion cost. An inter and intra-zonal congestion management scheme based on adjustment of schedules of generation and load has been discussed by Gribik et al. [49]. The schedule adjustments are based on nodal and marginal pricing. An optimal dispatch based congestion management scheme in a market having bilateral and multilateral transaction contracts is proposed in [50]. The various curtailment strategies in order to alleviate congestion are presented by authors. A factor known as willingness to pay to avoid curtailment is introduced to mitigate congestion effectively.

An AC-OPF based congestion management scheme is proposed in [51] .The congestion alleviation is obtained by optimal rescheduling of generation with minimum congestion relief cost. A sensitivity based approach using AC-OPF is discussed in [52]. Sensitivities of line flow to changes in generation is presented to mitigate congestion. However the authors have not made any effort to reduce the number of participating generators. Similar method for congestion management based on generation rescheduling has been discussed in [53] wherein a multi-objective particle swarm optimization is presented to find the optimal solution. The authors have utilized generator sensitivity for selection of participating generators. A similar approach has been adopted in [54] to mitigate congestion using generation rescheduling. Sivkumar and Devaraj [55] proposed the generation rescheduling based congestion management approach using Genetic algorithm. Nesamalar et al. [56] proposed an energy management scheme for congestion management of transmission system by rescheduling of generators using Cuckoo search algorithm. Different optimization algorithms for solving the problem of congestion management using generation rescheduling have been adopted by different authors.

The methods discussed above are utilized for different electricity market structure under deregulated environment.

2.4 Literature review on congestion management using FACTS devices

The management of congestion using traditional dispatch actions such as generation rescheduling, load curtailment is a non-cost free means of congestion alleviation as it involves the economic matters of electricity market. The up and down of generators' output directly affects the revenue generation of the market. Although they are easier in implementation and may still be needed in worst condition, but they are not preferable in competitive environment of electricity market as they are not transparent and may put a stop on the market from further development. Another method to manage congestion which does not touch the economic aspect of electricity market employs the use of FACTS devices. It is a cost free means of managing congestion as the marginal costs (and not the capital costs) involved in their usage are nominal.

FACTS devices reduce the power flows in heavily loaded lines which results in an increased loadability, low system losses, improved network stability, reduced cost of production and controlled power flow in the network with the fulfillment of constraints [57]-[58]. In deregulated electricity market, congestion management using FACTS devices is one of the widely used methods to alleviate congestion.

Reddy et al [59] has proposed a method to manage congestion using FACTS devices using TCSC and UPFC. In [60], a method to manage congestion in bilateral transaction model of electricity market has been presented. The method is compared with load curtailment method of congestion alleviation. Tiwari and Sood [61] has investigated the impact of FACTS device in pool and bilateral market on real power loss.and optimize the social welfare maximization problem. Choi and Moon [62] relieved the congestion in transmission line based on line flow sensitivity to FACTS devices. The line flow sensitivity is determined using dc load flow model. A supervisory controller based optimal power flow model with multiple objectives to avoid congestion, provide secure transmission and minimize active power losses has been derived in [63]. Skokjev et al [64] has discussed additional benefits extended by FACTS devices apart from congestion mitigation. An OPF based congestion management scheme using FACTS device has been proposed in [65]. The problem is optimized using evolutionary programming.

Although implementation of FACTS devices provide an effective solution to congestion management problems, there are certain issues related to it which require greater concern such as its optimal location, type, quantity, best allocation, parameter setting and installation cost [66]-[67]. High cost of FACTS devices make it necessary to utilize them effectively by optimally locating them in a power network so that maximum benefit with minimum expenditure can be achieved. The solution to congestion management problem is accomplished by placing it optimally in the network. In [68], Singh and David developed a simple and efficient model to locate FACTS devices optimally which is used to manage congestion by optimally controlling their parameters while Gerbex et al [69] locate the multi-type of FACTS devices (TCSC,TCPST, TCVR and SVC) optimally using genetic algorithm. Besides optimal location, the ratings and investment costs of FACTS devices are also optimized in [70]. The sensitivity factor based methods to find best location for FACTS devices has been widely adopted [71]–[73]. Singh and David [74] proposed overload sensitivity factor to find optimal location of TCSC and TCPAR for congestion management. An optimal location of FACTS device based on loss sensitivity factor is discussed in [75]. Line outage sensitivity factors are used by Hashemzadeh and Hosseini [76] to find the suitable locations of FACTS device. The authors have employed particle swarm optimization technique to optimize the FACTS device location for congestion management. However the sensitivity based method may not capture the non-linearity associated with the system. To avoid such condition Acharya and Mithulananthan [77] have proposed an LMP and congestion rent based method to find the optimal location of series FACTS devices in pool market. Similar to sensitivity based method they formed the priority list of potential locations of FACTS devices and select the location which is in top of the priority list. Such a priority list has also been proposed in [78] to determine the location of TCSC so that total congestion rent and total generation cost is minimum. An LMP based method for optimal allocation and size of series FACTS devices is also proposed in [79]. Taher [80] has proposed a method to optimally locate TCSC based on real power performance index and reduction of total system VAR losses. The location as well as size of FACTS devices are optimized using Simulated annealing in [81] to alleviate congestion in power system. Wibowo et al. [82] placed the FACTS devices optimally for congestion relief as well as voltage stability. The optimal placement of parallel i.e STATCOM and series FACTS devices i.e SSSC for congestion relief in deregulated environment is studied in [83]. Moreover, the optimal number of each FACTS device for congestion management is also investigated. A fuzzy based genetic algorithm for optimal placement and sizing of TCSC and SSSC in double sided auction market is studied in [84] while location and size of FACTS devices using particle swarm optimization algorithm is investigated in [85]. Ravi and Rajaram [86] have used improved particle swarm optimization algorithm for optimal placement and sizing of

STATCOM. A hybrid algorithm having combination of bacterial foraging with Nelder-Mead method is employed in [87] to find the optimal location and size of TCSC.

2.5 Literature review on congestion management using distributed generations

In new era of competitive electricity market, demand side approach for congestion management are getting more attention as it mitigates congestion more effectively and efficiently thereby improving the reliability and security of the power system [88]. Since distributed generations (DGs) [89]-[90] can be generally located in load pockets as negative power demand and also can respond quickly to the changing conditions of competitive electricity market, therefore these are getting an augmented interest in restructured power system operation and planning. Its strategical location and operation in system reduce losses, improve voltage profile, defer system upgrades and improve reliability of the system [91]–[97]. Also it is easy to install and simple to operate. With all these benefits, DGs can be considered as a powerful tool for demand side based congestion management in restructured power system [98].

The concept of congestion management of transmission lines through DGs was first introduced by Liu et al [99], who have utilized DG to manage the operations of power system. An optimal power flow based method is proposed in [100]-[101] in which authors have used DGs as a tool for congestion management. In [102], DG is employed for congestion management in a competitive electricity market having pool as well as bilateral transactions. Economical factors are taken into account for congestion management using DGs. A cost/benefit analysis approach has been proposed for implementation of DGs.

Similar to FACTS devices, the optimal placement and sizing of DG plays a vital role to achieve power system economics. The placement and size should be optimal such that maximum benefit could be generated with its implementation in the network otherwise in some situations it may jeopardize the power system operation and performance. An examination of placement and size of DGs have been discussed in [103]. Gautam and Mithulananthan have proposed an LMP based approach in [104]-[105] to find the optimal location of DG for congestion management in deregulated electricity market. The authors have proposed consumer payment as a tool to identify potential locations of DG allocation. A DC-OPF based congestion management method using DG to is presented in [106]. The authors have used LMP for optimal allocation and penetration of DGs. In [107], an AC-OPF is presented for congestion management. The optimal placement and output power level of DGs are evaluated based on LMP and congestion rent. The authors have proposed a performance index based on congestion rent to select the optimal size of DGs. A hybrid method employing OPF and PSO has been proposed in [108] to get the optimal size and site of DGs. Nabavi et al [109] have implemented PSO algorithm to get the optimal location and size of DGs to manage congestion in double sided auction market. In [110], Genetic algorithm is used to find optimal location and size of DGs. A sensitivity based method has been proposed by Singh and Parida [111] to allocate DGs for congestion relief as well as to achieve voltage security of the system. The load buses are ranked based on the sensitivity of the overloaded lines with respect to the bus injections for allocation of DGs. Genetic Algorithm is used to compute the new penetration level for DGs connected to these load buses. A method, independent of slack bus location, has been discussed in [112] to optimally place DGs for congestion management in deregulated power system. The optimal location of DG is evaluated using contribution factors based on bus impedance matrix. The bus impedance matrix is independent of slack bus location which complies with prevailing competitive environment of electricity market.

2.6 Conclusion

In the present chapter, a comprehensive review of the progress in the area of congestion management using generation rescheduling, FACTS devices and Distributed Generations has been presented. Also, different market structures have been reviewed. It is observed in the literature that before deregulation, congestion management was simpler and was carried out by optimal power flow using linear/ non-linear/ integer programming. With the advent of deregulation, management of congestion became more complex due to unbundled structure and presence of large number of independent market players. So, congestion management did not remain only technical issue, rather it became a complex combination of various economic and technical factors which require an application of meta-heuristic techniques such as Genetic Algorithm, Particle Swam Optimization, Ant Colony Optimization etc. to solve it.

It is observed from literature that different types of electricity market structure have been adopted by different authors. The suitability of a specific market structure depends on certain issues of country such as topographical and economic issues. Hybrid market model consisting of pool and bilateral transaction is adopted in present scenario of competitive electricity market. However, bilateral transactions may risk the security of power system.

It is also observed that traditional dispatch methods for congestion management such as load curtailment, generation re-dispatch etc. are overtaken by market oriented methods of congestion management. Although generation rescheduling method of congestion management has been widely used for congestion management in deregulated electricity market, it should be judiciously applied as improper rescheduling of generation may take away all the benefits of competitive electricity market. Sensitivity based methods are generally adopted to reschedule the generation for congestion management. Very less attention has been paid to schedule the generation on the basis of bids submitted by different generating utilities.

Besides generation rescheduling, application of FACTS devices is another means of congestion management which is widely used. However, selection of its location and size requires a great concern as its installation involves a heavy capital investment. Several techniques have been reported in literature for optimal placement and size of FACTS devices for congestion management. Even though it's appropriate placement and size selection remains an unsolved issue.

Implementation of DGs for congestion management is another method to relieve congestion. It is observed that DGs are extensively used to improve system's indicator such as voltage profile, power quality etc. The DGs are very powerful tool to manage distribution system. But from literature, it can be observed that it is seldom used for congestion management. Although it provides a good measure to manage transmission line congestion from demand side, it gets very less attention due to its high capital cost. Therefore, its location and penetration level presents a formidable challenge so that the investment involved in it is justified.

Further observations from literature reveal that several optimization techniques are available for solving the congestion management problem among which PSO is widely used meta-heuristic optimization techniques. The convergence of PSO algorithm depends upon the appropriate selection of its parameter value which is an area of great concern.

CHAPTER – 3

COMPETITIVE ELECTRICITY MARKET: OVERVIEW AND DESIGN

3.1 Introduction

In last two decades, electricity market has undergone major changes around the world. Although these changes are diverse in nature, their core aim is to develop towards a more competitive and open environment. This leads to a competitive electricity market where electricity is being traded as a commodity and the price of electricity is driven by market forces [12]-[15]. A competitive electricity market is one in which a number of suppliers (generators) compete with each other to sell their electricity to a number of competing customers (loads).

Competition is the foremost goal of electricity industry restructuring. From an economist's point of view, it is always desirable to have a market structure having perfect competition which is characterized generally by a condition where all firms in the industry are price takers and have freedom to enter and exit from the industry. These are identified based on three different criterion i.e independent, product substitute and entry criterion. However, it is rare to find these criteria get satisfied in any real market which often operates at a suboptimal level.

Competitive electricity market involves a very complex task to design it as it not only gets influenced by economic and engineering considerations but also by historical, political and social constraints. This complex problem of electricity market design does not have unique solution and differs in each country as well as among various regions. Although a large benefit is achieved by the society through competitive electricity market, there are some serious market design issues which need to be concerned. Therefore, each country should adopt a competitive market design such that it suits their economic, political and social environment and provide maximum benefit to the society.

3.2 Market structure and operation

3.2.1 Objectives of market operation

Secure and economic operation of power system are the two objectives behind establishing an electricity market. Whether it is a regulated or deregulated electricity market, secure operation of power system is of prime concern. In deregulated environment, this could be achieved through various services offered to the market. Reduction in electricity utilization cost is accomplished through economical operation of electricity market which is the prime reason for restructuring and also a means to improve power system security through its economics. This is achieved by adopting suitable market designs based on requirements of power system.

3.2.2 Electricity market models

The electricity market structure, whether existing or developing, is very much non-uniform due to variation in regional characteristics [16]-[18]. Some adopts centralized day-ahead and hour ahead markets for wholesale trading and a real-time market for balancing while others only select one or two centralized markets. Besides this, others may only offer bilateral contracts among participants without any centralized markets. To facilitate goals of electricity market, its several models have been considered which are outlined as follows.

a) PoolCo model

A PoolCo is based on the principle of open and transparent access of transmission line to all market participants [19]-[22]. It is a centralized market place where market for buyers and sellers is cleared by an Independent System Operator (ISO) who is responsible to maintain the secure operation of power system. A pool bidding or selling transaction is a price-quantity-based offer to sell into the pool by a Generation Company (GENCO) or to buy from pool by a Distribution Company (DISCO). The bids for trading of power in the electricity market are submitted to the pool by the GENCOS and DISCOS. The GENCOS would compete among themselves to get the right to supply energy to the grid without considering specific customers. If a GENCO participating in market, bids too high for their amount of power supply, it may not be able to sell. Conversely, DISCOS compete for right to purchase power from the grid and if their bids are too low as compared to others they are not able to buy power. The generators with low cost are rewarded in this market model. ISO, after evaluating all the bids, would produce a single electricity price (spot price) by implementing economic dispatch which would give a clear signal to the participants for taking decisions regarding consumption and investment. The spot price would be driven by market dynamics to a competitive level equal to the marginal cost of most efficient bidders. Spot price equal to the highest bid of winners are paid to the winning bidders in this market model.

b) Bilateral contracts model

Another model of an electricity market consists of bilateral contracts which are nothing but an agreement between seller and buyer of energy [25]-[30]. The traders negotiate the delivery and receipt of power and set the agreement terms and conditions without the involvement of ISO. The ISO only performs the task of keeping the system secure which would be accomplished by verifying the availability of sufficient transmission capacity to carry out the bilateral transactions agreed between traders. As the contract terms are specified by the traders, this form of electricity market model is very flexible. However, it suffers from disadvantages of high cost incurred due to contract negotiation and writing. This model has also the risk of creditworthiness of counterparties.

c)   Hybrid model

The features of both PoolCo as well as bilateral contract model of electricity market are combined to form a hybrid market [32]-[36]. The obligation to utilize PoolCo is eliminated in hybrid model. The customers are free to negotiate directly with suppliers for power supply agreement or they could choose to participate in pool bidding to sell/buy power at spot price. The market participants (sellers and buyers) who do not go under a bilateral contracts agreement would be served by the PoolCo. However, a true customer choice would be offered by allowing negotiation of power purchase agreements between suppliers and buyers. This would create a numerous pricing options as well as services to best meet individual customer needs.

3.2.3 Competitive market design

In general, electricity markets restructuring emphasize on functionally unbundle the vertically integrated system. The usual separation of vertically integrated system into generation, transmission and distribution is insufficient. The pool dispatch and the transmission system are two distinct but essential facilities. Although the bilateral market offers a choice to customers for energy sell and purchase, there are certain system limitations which make it difficult to achieve an efficient large scale bilateral electricity market. These barriers could be easily overcome by a system operator coordinated pool based electricity market [37].

The central requirement for the operation of a system is reliability and security [38]-[39]. With the advancement of technology and increase in demand, open access of transmission system requires a system operator to coordinate the transmission system usage. The system operator should operate the transmission system independently of the electrical utilities and should provide transparent and fair access to all market participants. It does not compete in the energy market but performs an essential task of maintaining the security and reliability of the system. Therefore, for an electricity market to be competitive, the spotlight is on the design of interaction between transmission and dispatch in terms of pricing and procedures. Such an interaction is shown in Figure 3.1 where the dispatch is managed centrally in competitive electricity market. For a competitive electricity market to operate efficiently, it is likely to differentiate short-run operations with long-run decisions. The system operator coordinates the short-run operations while long-run decisions involve investment for future expansion of transmission system. In competitive market, the participants include generators and customers where generators are price takers. The system with short-run operation is simple and once its economics get established, the requirement of long-run comes out to be more transparent.

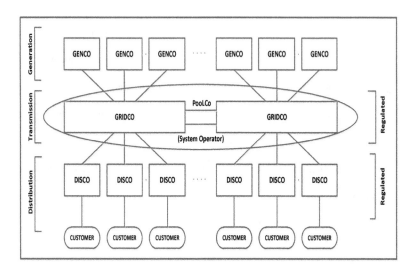

Figure 3.1 Competitive electricity market structure

3.3 Power transaction management issues

The motivation behind deregulation can be easily understood by the benefits achieved through it. Deregulation of electricity sector is as applicable as industry producing common goods with the approximation that there is little difference between goods and electrical energy. However, there are certain characteristics of electrical energy which need special consideration. These include:

(i) Inability to store electrical energy,

(ii) Large variation in daily and seasonal demand,

(iii) Power system operational requirements for reliability and security,

(iv) Network externalities,

(v) Limitations of transmission system to accommodate all the transactions

All these characteristics of electrical energy put together a challenge to the benefits promised by deregulation of an industry. The transactions of power in a competitive market give rise to certain issues which are as follows [40]-[43]:

a) Power flow on transmission networks

Transmission network operation is closely linked with the economy of an electricity market. Transmission network in electricity market is similar to a transportation network in an industry. However, unlike transportation network which is free to choose alternate paths for the movement of goods in order to achieve system economy, the transmission network has extremely limited paths to control the power flow. Also, unlike transportation network which assumes that goods when not moved can be stored, the transmission system does not have this characteristic as the electrical energy cannot be stored. The amount of flow of energy from one node to other is being strictly governed by physics of power system. Although the power system apparatus such as FACTS devices and phase shifting transformers can control the flow of power over an individual line, they involve a huge investment.

Every transmission system link has its limit on the amount of power transfer. These limits must be set to include normal operation as well as contingencies so that the power system continues to operate even after the occurrence of such contingencies [40]. The operation of power system within its limit is important for system security and reliability otherwise severe economical and social consequences can take place.

b) Congestion management

When production and consumption of electrical energy that would cause the operation of transmission system at or beyond one or more transfer limits, the system is said to be congested. It may also be defined as the inability of a transmission system to accommodate all the transactions. The control strategies applied for the operation of the transmission system below its transfer limits is known as congestion management. The term congestion got associated with power systems with deregulation, although it was present in vertically integrated power system. Before deregulation, the congestion is dealt in terms of steady-state security which is maintained by controlling the generator output. A single entity (vertically integrated utility), which controls generation, transmission and distribution, is responsible for the management of power flow within an area. The entity would try to achieve economy with lower generation cost and in the process of accomplishing this if any unusual condition arise due to less secure operation, they are held responsible for that. There is one decision making entity within which security and economics conflict could be traded off.

In regulated power system, the management of congestion is somewhat easier as all the power system components are under control of a single utility. Mostly sale and purchase of energy were among adjacent utilities, so unless the utility agreed that the negotiated transactions were in their best interest for both economy and security, it would not go forward.

In deregulated power system, the management of congestion is somewhat complex as compared to regulated system. The transmission system operator has a challenge to lay down a set of rules for congestion management that would ensure adequate control over generators and consumers in order to maintain power system security and reliability with maximum market efficiency. The rules should be transparent so that the occurrence of a particular outcome can be clear to all market participants. Also it should be robust to avoid creation of market power by exploiting congestion as a means to increase the profit by many aggressive entities which would lead to an inefficient market. Besides these, the rules should be fair enough to reflect its effect on different market participants. The adoption of a congestion management strategy depends upon the form of electricity market and congestion management itself cannot be separated from market considerations [41]-[42].

c) Market economics

Market performance is measured in terms of its social welfare which is a combination of energy cost and energy benefit to society as measured by willingness of the society to pay for it. For energy demand having zero price elasticity, i.e. demand is assumed to be price independent, the social welfare comes out to be negative of the total price paid for energy.

The social welfare is maximum in a perfect market. However, a perfect market is not possible in real world. Therefore, real market has somewhat lower values of social welfare as compared to perfect market and its efficiency is measured in terms of difference in values of social welfare.

To achieve a perfect competition in electricity market, following conditions are required [42]:

(i) generators in large number must be available to produce electrical energy
(ii) each generator competes to maximize their profits;
(iii) each generator is a price taker i.e. market price cannot be changed by generator by changing their bid;
(iv) market prices are known to all generators;
(v) costless transmissions.

Arguably, there is no existence of any of these conditions in real market. For a generator to be price taker, it requires incremental cost bidding to maximize its profit. The bidding of generator other than its incremental cost to increase its profit by exploiting market imperfections is known as strategic bidding. Strategic bidding will make the generator to have market power, if it increases its profits. Market power also exists if any means other than lowering the generation cost increases the profit of a generator. Obviously, the market power leads to market inefficiency [42].

d) Transmission tariffs

Transmission tariff is the price charged to the market participants for transmission system usage. The issue of transmission tariff in power flow management has three aspects.

Transmission tariff can ensure to generate enough revenue so that it covers the transmission system operator cost. In cases where the transmissions system operator and transmission system owners are separate, then the revenue from tariff must cover the cost of both entities. The tariff should be such that it covers both the operating cost as well as future transmission construction. While operating cost is easily covered, the cost of covering future transmission expansion is somewhat a hard task.

Transmission tariffs can be efficiently utilized as a tool to hedge congestion. It should reflect price signals to users of transmission system in order to manage congestion operationally such that electricity market remains efficient [42]-[43].

Transmission tariffs can ensure to generate revenue to cover the physical phenomenon of transmission losses.

e) Transmission losses

Transmission losses are another issue of concern in power flow management [42]-[43]. It is of equal importance as other issues explained above. Despite the fact that it may be included in transmission tariffs, it can also be handled separately. A number of loss management approaches have been in practice. Although the effect of loss may seem to be small in comparison with other market inefficiency potential sources, it should definitely be taken care of as efficiently as possible.

3.4 Conclusion

In this chapter, different types of electricity market have been modelled. It has been observed that the pool dispatch and the transmission system are two distinct but essential facilities for reliable operation of electricity market. For an electricity market to be competitive, the spotlight is on the design of interaction between transmission and dispatch in terms of pricing and procedures. Although the bilateral market offers a choice to customers for trading of energy, there are certain system limitations which make it difficult to achieve an efficient large scale bilateral electricity market. These barriers could be easily overcome by a system operator coordinated pool based electricity market. For a competitive electricity market to operate efficiently, it is likely to differentiate system operator coordinated short-run operations with long-run decisions involving investment. However, the system with short-run operation is simple and once its economics get established, the requirement of long-run comes out to be more transparent. Therefore a short-run system operator coordinated pool based electricity market is designed.

It has also been observed that power transaction in competitive electricity market has certain issues, such as congestion, market economics, transmission tariff and losses; which need to be handled with great precaution such that it would not take away the benefits of restructuring. As congestion is central to the issue of transmission tariffs, losses as well as market economics, therefore congestion management remains the central issue to power transfer management in competitive environment of electricity market. Market participants do not bother about the reliability and security of the system. They participate with the only aim of maximising their profit. Without robust congestion management strategy, their actions can put the transmission system operation in stake. Therefore, interaction of congestion management with energy market economics should be carefully accomplished such market inefficiency cannot take away the benefits promised by deregulation to the society.

CHAPTER – 4

CONGESTION MANAGEMENT USING GENERATION RESCHEDULING

4.1 Introduction

The success of deregulation in other sectors such as communication and airlines motivated the deregulation of electrical industry. During the last two decades, the deregulation of electricity sector has been witnessed all over the world. This has resulted in change in electricity sector operation philosophy. The introduction of competition due to deregulation causes the cost based electricity to transform into price based market commodity. This increased competition reduces the net electricity cost as the price of electricity is driven by market forces. The competition in new liberalized market causes each independent generating utility to sell all their generated power to the consumers. Hence they try to accommodate all their generated power on transmission line which may cause violation of transmission line limits such as thermal limit, voltage limit, stability limit etc. and thus makes the transmission line congested. The transmission line congestion may lead to tripping of overloaded lines, power system instability etc. and obviously increase the electricity cost as it causes the power system to deviate from its optimal operation. Hence the congestion needs to be alleviated as soon as possible.

The problem of management of congestion is considered as one of the fundamental problem of transmission management. However, it needs technical and financial tools for managing congestion in deregulated electricity market as compared to vertically integrated utility. The methods of congestion management generally include generation rescheduling, load curtailment, implementation of FACTS devices, etc.

However, the adoption of a congestion management technique depends on market structure. A specific congestion management technique suits a particular market. In Swiss market, an optimal power flow based re-dispatch method is used to mitigate congestion [113]. A day ahead forecast for congestion is made to take the corrective actions by ISO. Market splitting and counter trading is used in Nordic countries to manage congestion [114] while Locational Marginal Price (LMP) based congestion management method is used in California [115]. U.K. employs a combination of variety of congestion management methods for transmission line management. In this chapter, the problem of congestion management is addressed using real power generation rescheduling of generators.

4.2. Congestion management using generation rescheduling

One of the most widely used method for managing congestion around the world is by rescheduling the power output of generators. This method along with load curtailment is in use for a very long time. This method is not only adopted in deregulated market structure, but was also in use in vertically integrated system. A number of strategies have been adopted by different authors in order to reschedule the generators output for management of congestion efficiently. Globally, OPF based congestion management schemes with re-dispatch of generation and load curtailment has been extensively utilized to relieve the congestion [42]-[54].

A relative electrical distance based active power rescheduling for congestion management was discussed in [116]. Congestion management using optimal transaction by load curtailment in hybrid electricity market was discussed in [117]. A transmission congestion relief index is utilized in [118] to manage the generation rescheduling for congestion alleviation. In [119] generator sensitivity to the congested line is utilized to select the generators for participation in congestion management and the rescheduling is done accordingly whereas a similar approach is utilised in [120] to reschedule the generation and load shedding of buses sensitive to the congested line. Most of the above discussed methods make use of sensitivity factor to re-dispatch the generators but do not consider the economic aspects of competitive electricity market. In this chapter, an OPF based congestion management scheme using generation rescheduling is proposed. The proposed method finds the optimal selection of participating generators for rescheduling based on generator sensitivity factor considering the economic aspects of electricity market. Furthermore, the amount of rescheduling of the power output of participating generators depends upon the generator sensitivity as well as the rescheduling bids submitted by the generators to the system operator for regulating their output.

4.2.1 Generator sensitivity

The generators in power system are sensitive to the flow of power in transmission lines and they accordingly respond to the changes occurring in transmission network. However, some are more sensitive and some are less sensitive to these changes in transmission network. To find the effect of these changes on generators, a term generator sensitivity factor is introduced [121] which is defined as the change in active power flow on line due to change of active power generation of that generator. The generator sensitivity (GS) of generator-i to the flow of power on line l connected between bus-p and bus-q is given as [121].

$$GS_i^{pq} = \frac{\Delta P_{pq}}{\Delta P_{gi}} \tag{4.1}$$

where, P_{pq} denotes the active power flow on congested line-l connected between bus-p and bus-q and P_{gi} is the active power generation of generator-i.

4.2.2 Problem formulation for congestion management

With deregulation of power sector, the problem of managing congestion has somewhat become more complex as compared to that in vertically integrated utility. Existence of competition in electricity market further makes it more difficult to manage congestion. The present work considers a congestion management scheme in pool market structure wherein generators submits bid to serve the load. The congestion management problem is formulated as minimization of active power rescheduling cost of generators which is given as [121].

$$Minimize \sum_{i}^{n_g} RC_i(\Delta P_{gi}) . \Delta P_{gi} \qquad (4.2)$$

where, $\Delta P_{gi} = f(GS, \Delta P_{bid})$

subject to following constraints:

1. Power balance equality constraint

$$\sum_{i=1}^{n_g} \Delta P_{gi} = 0 \qquad (4.3)$$

2. Operating limit inequality constraint

$$\Delta P_{gi}^{min} \leq \Delta P_{gi} \leq \Delta P_{gi}^{max} \; ; \quad i = 1, 2, \dots n_g \qquad (4.4)$$

where,

$$\Delta P_{gi}^{min} = P_{gi} - P_{gi}^{min}$$

$$\Delta P_{gi}^{max} = P_{gi}^{max} - P_{gi}$$

3. Line flow inequality constraint

$$\sum_{i=1}^{N} (GS_{gi}^{pq} . \Delta P_{gi}) + F_L \leq F_L^{max} \; ; \quad L = 1, 2, \dots n_L \qquad (4.5)$$

where, RC_i is the rescheduling cost of generator at bus-i,

ΔP_{gi} is the active power adjustment of generator at bus-i,

ΔP_{bid} is the bid submitted by different generators for their active power adjustment,

ΔP_{gi}^{min} and ΔP_{gi}^{max} are respectively the minimum and maximum limit of active power adjustments of genearator at bus-i

GS_{gi}^{pq} is the generator sensitivity of generator at bus-i,

F_L represents the power flow on line-l considering all the contracts,

F_L^{max} denotes the line flow limit of line-l connected between bus-p and bus-q,

n_L represents the total number of lines in the system.

4.2.3 Selection of participating generators for congestion management

The generators participating for congestion management are selected based on their calculated GS values given by equation (4.1).

After neglecting the coupling between active power and voltage i.e P-V coupling, equation (4.1) can be written as

$$GS_i = \frac{\partial P_{pq}}{\partial \theta_p} \cdot \frac{\partial \theta_p}{\partial P_{gi}} + \frac{\partial P_{pq}}{\partial \theta_q} \cdot \frac{\partial \theta_q}{\partial P_{gi}} \qquad (4.6)$$

The power flow equation for a congested line can be given as

$$P_{pq} = -V_p^2 G_{pq} + V_p V_q G_{pq} \cos(\theta_p - \theta_q) + V_p V_q B_{pq} \sin(\theta_p - \theta_q) \qquad (4.7)$$

Differentiating equation (4.7) with respect to θ_p and θ_q we get,

$$\frac{\partial P_{pq}}{\partial \theta_p} = -V_p V_q G_{pq} \sin(\theta_p - \theta_q) + V_p V_q B_{pq} \cos(\theta_p - \theta_q) \qquad (4.8)$$

$$\frac{\partial P_{pq}}{\partial \theta_q} = V_p V_q G_{pq} \sin(\theta_p - \theta_q) + V_p V_q B_{pq} \cos(\theta_p - \theta_q) = -\frac{\partial P_{pq}}{\partial \theta_p} \qquad (4.9)$$

The active power injected at any bus-s of the system can be expressed with equation (4.10).

$$P_s = |V_s| \sum_{k=1}^{n} ((G_{sk} \cos(\theta_s - \theta_k) + B_{sk} \sin(\theta_s - \theta_k)) |V_k|) \qquad (4.10)$$

$$= |V_s|^2 G_{ss} + |V_s| \sum_{\substack{k=1 \\ k \neq s}}^{n} ((G_{sk} \cos(\theta_s - \theta_k) + B_{sk} \sin(\theta_s - \theta_k)) |V_k|) \qquad (4.11)$$

Differentiating equation (4.11) with respect to θ_s and θ_k we get,

$$\frac{\partial P_s}{\partial \theta_k} = |V_s||V_k|(G_{sk} \sin(\theta_s - \theta_k) - B_{sk} \cos(\theta_s - \theta_k)) \qquad (4.12)$$

$$\frac{\partial P_s}{\partial \theta_s} = |V_s| \sum_{\substack{k=1 \\ k \neq s}}^{n} ((-G_{sk} \sin(\theta_s - \theta_k) + B_{sk} \cos(\theta_s - \theta_k)) |V_k|) \qquad (4.13)$$

If we neglect the P-V coupling, the relation between the change in active power and voltage phase angles at system buses can be expressed in matrix form as

$$[\Delta P]_{n \times 1} = [H]_{n \times n} [\Delta \theta]_{n \times 1} \qquad (4.14)$$

where,

$$[H]_{n \times n} = \begin{bmatrix} \frac{\partial P_1}{\partial \theta_1} & \frac{\partial P_1}{\partial \theta_2} & \cdots & \frac{\partial P_1}{\partial \theta_n} \\ \frac{\partial P_2}{\partial \theta_1} & \frac{\partial P_2}{\partial \theta_2} & \cdots & \frac{\partial P_2}{\partial \theta_n} \\ \vdots & \vdots & & \vdots \\ \frac{\partial P_n}{\partial \theta_1} & \frac{\partial P_n}{\partial \theta_2} & \cdots & \frac{\partial P_n}{\partial \theta_n} \end{bmatrix} \qquad (4.15)$$

Therefore,

$$[\Delta \theta] = [H]^{-1} [\Delta P] \qquad (4.16)$$

If $\quad [M] = [H]^{-1} \qquad (4.17)$

Then,

$$[\Delta \theta] = [M][\Delta P] \qquad (4.18)$$

Considering bus-1 as reference bus, the matrix [M] can be modified by eliminating row-1 and column-1 corresponding to reference bus as given by equation (4.19).

$$[\Delta\theta]_{n\times 1} = \begin{bmatrix} 0 & 0 \\ 0 & [M_{-1}] \end{bmatrix}_{n\times n} [\Delta P]_{n\times 1} \qquad (4.19)$$

The modified [M] gives the values of terms $\partial\theta_p/\partial P_{gi}$ and $\partial\theta_q/\partial P_{gi}$ of equation (4.6) in order to calculate the GS values of all generators. Generators having large GS values are selected to participate in congestion management and hence reschedule their generation output as they have greater influence to the flow of power on congested line.

The amount of rescheduling of generators power output and consequently the rescheduling cost of generators are optimized by particle swarm optimization algorithm which is discussed in further section.

4.2.4 Particle swarm optimization (PSO)

PSO is an efficient and promising optimization technique used for non-convex optimization problems. It was first proposed by Kennedy and Eberhart in 1995 [122]. It is a population based optimization algorithm which is motivated by social and cooperative behaviour of organisms such as fish, birds etc. It consists of a population of potential solution called particles. Each particles search for a potential solution in multi-dimensional search space and update its position and velocity from time to time according to previous experience of its own and its neighbours. In a z-dimensional search space, the position and velocity of particle-n along with best position of an individual particle (position best) and a collective best position among all particles in population (global best) are represented in matrix form by equation (4.20).

$$\begin{bmatrix} X_n \\ V_n \\ P_n \\ G_b \end{bmatrix} = \begin{bmatrix} x_{n1} & x_{n2} & x_{n3} & \cdots\cdots & x_{nz} \\ v_{n1} & v_{n1} & v_{n1} & \cdots\cdots & v_{nz} \\ P_{n1} & P_{n1} & P_{n1} & \cdots\cdots & P_{nz} \\ G_{b1} & G_{b1} & G_{b1} & \cdots\cdots & G_{bz} \end{bmatrix} \qquad (4.20)$$

Both the information of particle best position and global best position are used by the particle to update their position and velocities as given by equations (4.21) and (4.22) respectively. The particles continue searching for solution until a convergence criteria or maximum iterations is achieved.

$$V_n^{t+1} = wV_n^t + c_1.r_1.(P_n^t - X_n^t) + c_2.r_2.(G_b^t - X_n^t) \qquad (4.21)$$

$$X_n^{t+1} = X_n^t + V_n^{t+1} \qquad (4.22)$$

where, w is a positive value called inertia weight and is given as:

$$w = w_{max} - (w_{max} - w_{min})\frac{t}{T_{max}}$$

r_1 and r_2 are random values between 0 and 1

c_1 and c_2 are called acceleration coefficients and $(c_1 + c_2) \geq 4.0$.

PSO has found its implementation in optimization of various power system problems. It has been enthusiastically used in solving congestion management problem as evident from literature review. It has been successfully implemented to manage congestion by generation rescheduling [123]-[125]. PSO technique was utilized in [123] to relieve congestion using generator sensitivity factor and thereby minimizing the total cost of generation. Similarly in [124] optimal numbers of generators for congestion management were selected based on generator sensitivities to the flow of power on congested line and PSO was used to reschedule the selected generators for congestion relief. An OPF based congestion management scheme using fitness distance ratio particle swarm optimization was discussed in [125].

4.2.5 Congestion management algorithm using PSO

The algorithm adopted for congestion management using PSO can be illustrated by the flowchart shown in Figure 4.1.

The steps adopted for PSO optimization algorithm to find the optimal solution of the objective function given by equation (4.2) with binding constraints given by equations (4.3) to (4.5) are explained as follows:

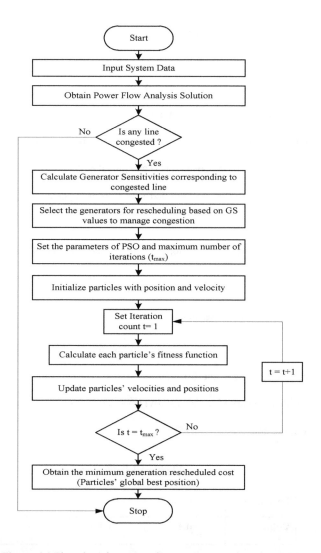

Figure 4.1 Flowchart for congestion management based on PSO

Step 1: The particles are generated and initialized with position and velocity. Every particle will have z-dimensions, z being the number of generators participating in congestion management, and the value of z variables denotes the amount of power rescheduling required by generators in order to relieve congestion.

Step 2: The binding equality constraint given by equation (4.3) and inequality constraints given by equation (4.4) and (4.5) are tested individually based on system states represented by an individual particle. If the particle does not satisfy any of the constraint, then it is regenerated.

Step 3: The optimal objective fitness values for every particle are calculated to determine the position best and global best values.

Step 4: The particles' position and velocities are updated using equation (4.21) and (4.22) respectively.

Step 5: If the pre-specified stop criterion or maximum number of iterations specified are reached, the optimization program is stopped, otherwise go to step 2.

4.3 Results and discussions

The performance of the proposed algorithm for congestion management is tested on IEEE 14-bus system and IEEE 30-bus system. The simulation studies of the proposed algorithm are carried out using MATLAB. The various parameters taken for PSO are given in Table 4.1 [124]-[125].

Table 4.1 PSO parameters

Parameters	w_{min}	w_{max}	c_1	c_2
Values	0.4	0.9	2.0	2.0

4.3.1 Results for IEEE 14-bus system

The power flow solution for IEEE 14-bus system is shown in Table 4.2 which gives that the power flow in each line is within its thermal limit except the line connected between bus-2 and bus-3. The real power flow in this line is 35.48 MW which is above its thermal limit.

Table 4.2 Power flow results for IEEE 14-bus system

Line No.	From Bus	To Bus	P (MW)	P_{max} (MW)
1	1	2	92.38	120
2	1	5	51.39	65
3	**2**	**3**	**35.48**	**35**
4	2	4	40.19	65
5	2	5	33.52	50
6	3	4	-0.11	65
7	4	5	-29.09	45
8	4	7	30.74	55
9	4	9	17.56	32
10	5	6	46.20	55
11	6	11	8.90	18
12	6	12	8.01	32
13	6	13	18.08	32
14	7	8	0.00	32
15	7	9	30.74	35
16	9	10	9.85	32
17	9	14	8.95	32
18	10	11	-5.18	12
19	12	13	1.83	12
20	13	14	6.16	12

The GS values calculated for active power flow on congested line is shown in Table 4.3. A graphical representation of GS values calculated for IEEE 14-bus system is also shown in Figure 4.3. A negative GS value shows that the increase in generation of that generator will decrease the power flow on congested line while a positive GS value shows that the increase in generation of that generator will increase the power flow on the congested line for which it is calculated. The GS values calculated for generators of IEEE 14-bus system are utilized for calculating the amount of power re-dispatch and hence rescheduling cost using PSO with maximum iterations set as 500 and particle size is taken as 70.

A graphical representation of the power flow in different lines of IEEE 14-bus system is also illustrated in Figure 4.2.

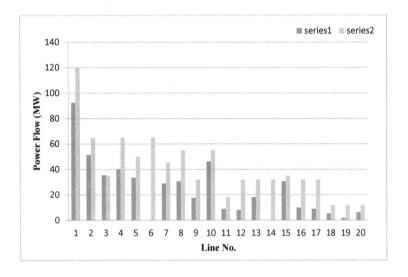

Figure 4.2 Power flow for IEEE 14-bus system before generation rescheduling

Series 1: Power flow in lines without generation rescheduling

Series 2: Power flow rating of lines

Table 4.3 Generator sensitivity values of IEEE 14-bus system for congested line connected between bus-2 and bus-3

Generator No.	1	2	3	6	8
GS Values	0	0.04397	-0.46111	-0.00983	-0.01264

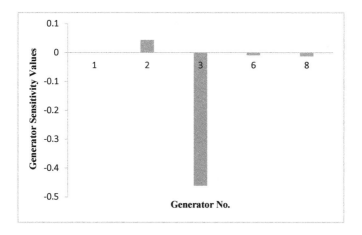

Figure 4.3 GS values for IEEE 14-bus system

Two different cases have been considered in order to analyse the effect of generator sensitivities as well as the bids submitted by the generators for regulating their output on the amount of rescheduling power output of generators and hence the generation rescheduling cost of the system. In case I, all the generators are considered to participate for congestion management while in case II generator at bus-6, due to its low generator sensitivity, does not participate in management of congestion. The slack bus generator i.e. generator at bus-1 will also participate in congestion management as it will take care of the losses in the system.

The bids taken for the calculation of generation rescheduling cost for up and down regulation of real power generation for different generators is shown in Table 4.4. The results thus obtained for two different cases is shown in Table 4.5 which depicts that both the active power rescheduling and the total rescheduling cost obtained is less in case II when the generator at bus-6 is not considered to participate in rescheduling for congestion management. Also, a considerable amount of active power rescheduling of generator at bus-6 in case I is taken up by generator at bus-8 in case II which has low value of bid for regulating up the output as compared to other generator having negative generator sensitivity.

Table 4.4 Rescheduling bids submitted by different generators of IEEE 14-bus system

Generator No.	1	2	3	6	8
Bids ($/MW^2-Day)	11	17	19	20	15

Table 4.5 Rescheduling results for IEEE 14-bus system

Generator No.	Active Power Rescheduling (MW)	
	Case I	Case II
ΔP_1	-17.3	-11.8
ΔP_2	-1.9	-6.3
ΔP_3	12.8	12.2
ΔP_6	3.9	Not participated
ΔP_8	2.5	5.9
Total ΔP	38.4	36.2
Cost ($/hr)	286.1	231.5

Table 4.6 and Table 4.7 show the load flow results for the above two cases and reveal that congestion is relived in both the cases considered for generation rescheduling.

A graphical representation of the power flow for the considered two cases is illustrated in Figure 4.4 and Figure 4.5 respectively which also reflects the mitigation of congestion from the system.

Table 4.6 Power flow results for IEEE 14-bus system for case I

Line No.	From Bus	To Bus	P (MW)	P_{max} (MW)
1	1	2	79.47	120
2	1	5	45.73	65
3	**2**	**3**	**27.33**	**35**
4	2	4	36.73	65
5	2	5	30.72	50
6	3	4	4.74	65
7	4	5	-26.30	45
8	4	7	29.56	55
9	4	9	17.42	32
10	5	6	40.92	55
11	6	11	8.22	18
12	6	12	7.77	32
13	6	13	17.63	32
14	7	8	-2.50	32
15	7	9	32.06	35
16	9	10	10.43	32
17	9	14	9.56	32
18	10	11	-4.62	12
19	12	13	1.60	12
20	13	14	5.51	12

Table 4.7 Power flow results for IEEE 14-bus system for case II

Line No.	From Bus	To Bus	P (MW)	P_{max} (MW)
1	1	2	83.92	120
2	1	5	46.92	65
3	**2**	**3**	**27.51**	**35**
4	2	4	36.50	65
5	2	5	30.70	50
6	3	4	4.32	65
7	4	5	-25.40	45
8	4	7	28.15	55
9	4	9	17.29	32
10	5	6	42.94	55
11	6	11	7.07	18
12	6	12	7.63	32
13	6	13	17.05	32
14	7	8	-5.90	32
15	7	9	34.05	35
16	9	10	11.56	32
17	9	14	10.28	32
18	10	11	-3.49	12
19	12	13	1.46	12
20	13	14	4.80	12

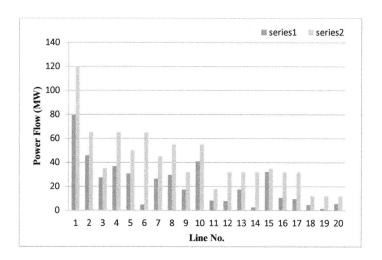

Figure 4.4 Power flow for IEEE 14-bus system after generation rescheduling (case I)

Series 1: Power flow in lines after generation rescheduling for case I

Series 2: Power flow rating of lines

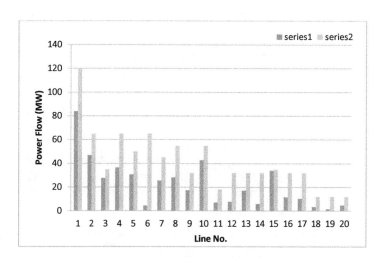

Figure 4.5 Power flow for IEEE 14-bus system after generation rescheduling (case II)

Series 1: Power flow in lines after generation rescheduling for case II

Series 2: Power flow rating of lines

Also to analyse the effect of change of bid of a generator for rescheduling their power output on amount of rescheduling and rescheduling cost, the bid of generator at bus- 3 is increased to $22/MW^2$-day. For this increase in bid of generator at bus- 3, the results are shown in Table 4.8 which reveals that output of generator at bus- 3 reduces to 11.9 MW as compared to 12.8 MW when its rescheduling bid was $19/MW^2$-day. This shows the increase or decrease of bid of a generator respectively decreases or increases its amount of rescheduling for congestion mitigation. Subsequently, all the generators participating in congestion management reschedule their power output with the change of bids of generator at bus-3. For this, the cost of generation rescheduling is found to be $293.6/hr while it is $306.5/hr if the generators do not reschedule their output after the change of bid. Although there is an increase in total amount of power rescheduling, there is a net saving in rescheduling cost when the generators output power rescheduling is made a function of generators bids along with generator sensitivity and is illustrated in Figure 4.6.

Table 4.8 Effect of change of bids on generation rescheduling for IEEE 14-bus system

Generator No.	Active Power Rescheduling (MW)		
	Before Change of Bid	After Change of Bid	
		Without Generation Rescheduling	With Generation Rescheduling
ΔP_1	-17.3	-17.3	-17.5
ΔP_2	-1.9	-1.9	-1.8
ΔP_3	12.8	12.8	11.9
ΔP_6	3.9	3.9	4.2
ΔP_8	2.5	2.5	3.2
Total ΔP	38.4	38.4	38.6
Cost ($/hr)	286.1	306.5	293.6

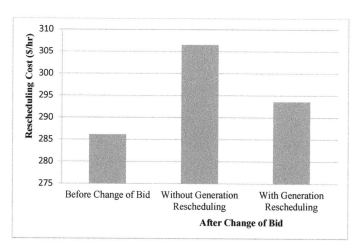

Figure 4.6 Effect of change of bid on generation rescheduling cost for IEEE 14-bus system

4.3.2 Results for IEEE 30-bus system

The power flow solution for IEEE 30-bus system is shown in Table 4.9 which gives that the line connected between bus-1 and bus-2 is congested as power flow in this line is above its thermal limit which is also illustrated in Figure 4.7.

Table 4.9 Power flow results for IEEE 30-bus system

Line No.	From Bus	To Bus	P (MW)	P_{max} (MW)
1	1	2	**170.14**	**130**
2	1	3	87.79	130
3	2	4	43.62	65
4	3	4	82.27	130
5	2	5	82.29	90
6	2	6	60.35	65
7	4	6	72.27	90
8	5	7	-14.85	70
9	6	7	38.20	130

Line No.	From Bus	To Bus	P (MW)	P_{max} (MW)
10	6	8	29.49	32
11	6	9	27.80	65
12	6	10	15.88	32
13	9	11	0.00	65
14	9	10	27.80	65
15	4	12	44.15	65
16	12	13	0.00	65
17	12	14	7.79	32
18	12	15	17.64	32
19	12	16	7.52	32
20	14	15	1.52	16
21	16	17	3.96	16
22	15	18	6.29	16
23	18	19	3.05	16
24	19	20	-6.46	32
25	10	20	8.75	32
26	10	17	5.07	32
27	10	21	18.29	32
28	10	22	5.78	32
29	21	23	0.64	32
30	15	23	4.45	16
31	22	24	5.75	16
32	23	24	1.86	16
33	24	25	-1.14	16
34	25	26	3.54	16
35	25	27	-4.69	16
36	28	27	18.00	32
37	27	29	6.19	16
38	27	30	7.09	16
39	29	30	3.70	16
40	8	28	-0.61	32
41	6	28	18.67	32

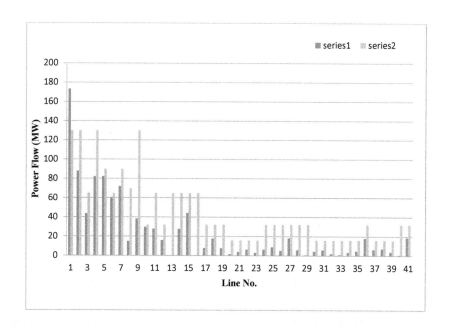

Figure 4.7 Power flow for IEEE 30-bus system before generation rescheduling

Series 1: Power flow rating of lines

Series 2: Power flow in lines before generation rescheduling

The GS values calculated for active power flow on congested line is shown in Table 4.10 which reveals that all the generators have high values of GS. Therefore all generators will take part in congestion management and hence will reschedule their generation. A graphical representation of GS values calculated for IEEE 30-bus system is also shown in Figure 4.8.

Table 4.10 Generator sensitivity values of IEEE 30-bus system for congested line connected between bus-1 and bus-2

Generator No.	1	2	5	8	11	13
GS Values	0	-0.8908	-0.8527	-0.7394	-0.7258	-0.6869

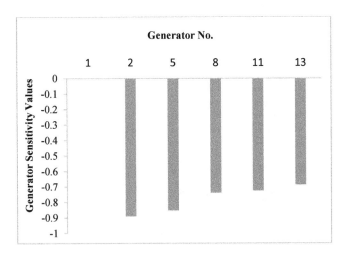

Figure 4.8 GS values for IEEE 30-bus system

The GS values calculated for generators along with rescheduling bids are utilized for calculating the amount of power re-dispatch and hence generation rescheduling cost. The optimization of rescheduling cost is done using PSO with all the parameters remaining same as taken for IEEE 14-bus system.

Since the generator sensitivities of all the generators are high, only one case is considered as contrast to that of IEEE-14 bus system. All the generators along with generator at slack bus (bus-1) will take part in congestion management and hence will reschedule their power generation in order to mitigate congestion.

The bids taken for the calculation of generation rescheduling cost for up and down regulation of real power generation for different generators of IEEE 30-bus system are shown in Table 4.11. The generation rescheduling result thus obtained is shown in Table 4.12 which depicts that the total active power rescheduling required for managing congestion is 94.8 MW and the total rescheduling cost obtained is $1472.3/hr.

Table 4.11 Rescheduling bids submitted by different generators of IEEE 30-bus system

Generator No.	1	2	5	8	11	13
Bids ($/MW^2-Day)	11	17	19	20	15	10

Table 4.12 Rescheduling results for IEEE 30-bus system

Generator No.	Active Power Rescheduling (MW)
ΔP_1	-47.4
ΔP_2	19.5
ΔP_5	12.4
ΔP_8	3.6
ΔP_{11}	7.0
ΔP_{13}	4.9
Total ΔP	94.8
Cost ($/hr)	1472.3

Table 4.13 shows the power flow results for IEEE 30-bus system after rescheduling of generators output for congestion management. It shows that the congestion on line-1 is effectively relieved by rescheduling the generator output by an amount of 94.8MW.

Table 4.13 Power flow results for IEEE 30-bus system after generation rescheduling

Line No.	From Bus	To Bus	P (MW)	P_{max} (MW)
1	1	2	125.87	130
2	1	3	72.71	130
3	2	4	41.69	65
4	3	4	68.17	130
5	2	5	73.32	90

Line No.	From Bus	To Bus	P (MW)	P_{max} (MW)
6	2	6	55.85	65
7	4	6	61.71	90
8	5	7	-10.83	70
9	6	7	34.02	130
10	6	8	26.24	32
11	6	9	22.95	65
12	6	10	14.49	32
13	9	11	-7.00	65
14	9	10	29.95	65
15	4	12	39.04	65
16	12	13	-4.90	65
17	12	14	7.77	32
18	12	15	17.56	32
19	12	16	7.41	32
20	14	15	1.50	16
21	16	17	3.86	16
22	15	18	6.23	16
23	18	19	2.99	16
24	19	20	-6.52	32
25	10	20	8.81	32
26	10	17	5.17	32
27	10	21	18.59	32
28	10	22	6.07	32
29	21	23	0.94	32
30	15	23	4.41	16
31	22	24	6.04	16
32	23	24	2.12	16
33	24	25	-0.59	16
34	25	26	3.54	16
35	25	27	-4.14	16
36	28	27	17.44	32
37	27	29	6.19	16
38	27	30	7.09	16
39	29	30	3.70	16
40	8	28	-0.24	32
41	6	28	17.73	32

A graphical representation of the power flow in different lines of IEEE 30-bus system is shown in Figure 4.9.

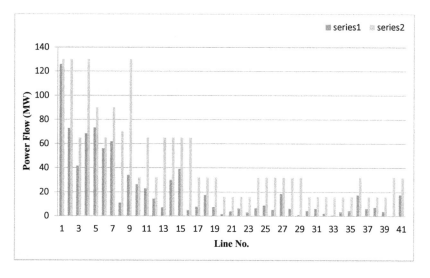

Figure 4.9 Line flow for IEEE 30-bus system after generation rescheduling

Series 1: Power flow rating of lines

Series 2: Power flow in lines after generation rescheduling

To analyse the effect of change of bid of a generator for rescheduling their power output on amount of rescheduling and rescheduling cost, the bid of generator at bus-5 is increased to $22/MW^2$-day. For this increase in bid, the results are shown in Table 4.14 which reveals that output of generator at bus-5 reduces to 11.1 MW as compared to 12.4 MW when its rescheduling bid was $19/MW^2$-day. Subsequently, all the generators participating in congestion management reschedule their power output with the change of bids of generator at bus-5. For this, the cost of generation rescheduling is found to be $1480.7/hr while it is $1491.5/hr if the generators do not reschedule their output after the change of bid. This shows that there is a net saving in the rescheduling cost when the generators output power rescheduling is made a function of rescheduling bids of generators along with generator sensitivity and is illustrated in Figure 4.10.

Table 4.14 Effect of change of bids on generation rescheduling for IEEE 30-bus system

Generator No.	Active Power Rescheduling (MW)		
	Before Change of Bid	After Change of Bid	
		Without Generation Rescheduling	With Generation Rescheduling
ΔP_1	-47.4	-47.4	-47.5
ΔP_2	19.5	19.5	20.4
ΔP_5	12.4	12.4	11.1
ΔP_8	3.6	3.6	3.5
ΔP_{11}	7.0	7.0	7.2
ΔP_{13}	4.9	4.9	5.3
Total ΔP	94.8	94.8	95.0
Cost ($/hr)	1472.3	1491.5	1480.7

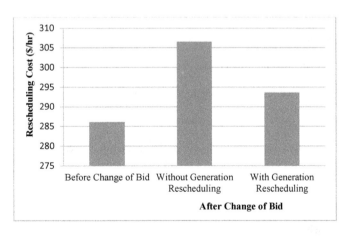

Figure 4.10 Effect of change of bid on generation rescheduling cost for IEEE 30-bus system

4.4 Conclusion

A congestion management methodology based on generation rescheduling has been proposed. The selection of participating generators is done on the basis of generator sensitivity values calculated for each generator for power flow on congested line. The amount of rescheduling of generators and hence the generation rescheduling cost is the function of generator sensitivity as well as the bids submitted by each generator for rescheduling their power output. The proposed algorithm has been tested on IEEE 14-bus system and IEEE 30-bus system.

It has been observed that the congestion management by optimal selection of generators based on their magnitude of generator sensitivities as well as generator's rescheduling bids gives more optimal generation rescheduling cost as compared to that when only generator sensitivities are taken into account for generators selection. Also, it has been found that the change of rescheduling bid of a generator accordingly changes its power generation for congestion management thereby providing the minimal rescheduling cost.

CHAPTER – 5

CONGESTION MANAGEMENT USING DISTRIBUTED GENERATION

5.1 Introduction

In new era of competitive electricity market, demand side approach for congestion management are getting more attention as it mitigates congestion more effectively and efficiently thereby improving the reliability and security of the power system [88]. Since distributed generations (DGs) can be generally located in load pockets as negative power demand and also can respond quickly to the changing conditions of competitive electricity market, therefore these are getting an augmented interest in restructured power system operation and planning. Its strategical location and operation in system reduce losses, improve voltage profile, defer system upgrades and improve reliability of the system [89]-[98]. Also it is easy to install and simple to operate. With all these benefits, DGs are being extensively used for congestion management in restructured power system [100]-[101]. Therefore, this chapter focuses on implementation of DGs for managing congestion by bifurcating the power system network into different congestion zones.

5.2 Zonal based congestion management using distributed generation

Since congestion zones identification in a system reduce the burden involved in computation of re-dispatch and transaction curtailments required for congestion alleviation, therefore feasibility of several zonal/cluster based congestion management techniques have been discussed in [126]-[128]. In [129]-[131], the selection of generators from most sensitive congestion zone for congestion management are achieved based on real and reactive power transmission congestion distribution factors. A similar approach for zonal congestion management has been discussed in [132] and [133]. In [133], real and reactive congestion distribution factors have been proposed for zonal congestion management has been proposed while ac transmission distribution factors have been proposed in [133]. The identification of different congestion zones or clusters have been by computing congestion distribution factor have been proposed in [134] in which type 1 cluster is the most sensitive.

In this chapter, a new zonal based congestion management approach in deregulated electricity market is proposed which is based on locational marginal price (LMP). Since LMP gives an economic signal [135] and the difference of LMP of buses across a line is measure of degree of congestion of that line, therefore it can be effectively and reliably utilized in deregulated electricity market [136]. With the identification of zone being most sensitive to congestion, the congestion is managed by optimally placing the distributed generation in that zone.

5.2.1 Locational marginal price (LMP)

LMP at a bus is defined as the marginal cost of supplying the next increment of electric energy at a specific bus while considering the generation marginal cost and the physical aspects of the transmission system [137]-[138]. It gives an economic signal to the electricity market and is therefore preferred these days in most of the electricity market to manage congestion. It consists of three components, marginal energy component which remains same for all buses, loss component and congestion component. Therefore, LMP at bus-m can be written as:

$$LMP_m = MEC_m + LC_m + CC_m \qquad (5.1)$$

where MEC$_m$ is marginal energy component, LC$_m$ is loss component and CC$_m$ is the congestion component of LMP at bus-m. Since the marginal energy cost remains same at all buses, therefore for small loss (negligible increase in loss) the LMP difference between two buses gives the congestion cost.

5.2.2 Problem formulation

To evaluate the nodal prices of electricity, the same problem which is formulated in previous chapter is considered. It is formulated as an optimal power flow (OPF) function in pool based deregulated electricity market, having no demand bid, with the objective of minimization of generation cost of electricity given by equation (5.2) while all other constraints are satisfied.

$$Minimize \sum_{k=1}^{n_g} C_i(P_{G_i}) \tag{5.2}$$

where n_g is the total number of generating units and $C_i(P_{Gi})$ is the cost of electricity generation of ith generating unit given as quadratic cost function:

$$C_i(P_{G_i}) = a_i \cdot (P_{G_i})^2 + b_i \cdot (P_{G_i}) + c_i \tag{5.3}$$

where a_i, b_i and c_i are the cost coefficients and P_{Gi} is the amount of electricity generation of ith unit.

The above objective functions are subjected to following constraints.

1. Power balance constraint at each node

$$P_m - P_{G_m} + P_{D_m} = 0; \quad m = 1,2,\ldots\ldots,n_b \tag{5.4}$$

$$Q_m - Q_{G_m} + Q_{D_m} = 0; \quad m = 1,2,\ldots\ldots,n_b \tag{5.5}$$

2. Generator operating limit constraint

$$P_{G_i}^{min} \leq P_{G_i} \leq P_{G_i}^{max}; \quad i = 1,2,\ldots,n_g \tag{5.6}$$

$$Q_{G_i}^{min} \leq Q_{G_i} \leq Q_{G_i}^{max}; \quad i = 1,2,\ldots,n_g \tag{5.7}$$

3. Line flow constraints

$$F_L \leq F_L^{max} \; ; \quad L = 1, 2, \ldots, n_L \tag{5.8}$$

4. Bus voltage limit

$$V_m^{min} \leq V_m \leq V_m^{max} \; ; \quad m = 1, 2, \ldots, n_b \tag{5.9}$$

where n_b is the total number of system buses, P_{Gk}^{min} and P_{Gk}^{max} are respectively the minimum and maximum real power output limits of kth generator, Q_{Gk}^{min} and Q_{Gk}^{max} are respectively the minimum and maximum reactive power output limits of kth generator, F_L denotes the flow of power on transmission line-L connected between bus-m and bus-n due to accommodation of all contracts, F_L^{max} is the power flow limit of line-L connected between bus m and bus-n, n_L denotes the total number of lines and V_m^{min} and V_m^{max} are respectively the minimum and maximum voltage limits at bus-m.

The optimization of the objective function incorporating all the constraints is done using Lagrangian method. The Lagrangian function of the optimization problem including all the constraints in objective function is written as:

$$\mathcal{L} = \sum_{k=1}^{n_g} C_i(P_{G_i}) + \sum_{m=1}^{n_b} \lambda_{P_m}(P_m - P_{G_m} + P_{D_m}) + \sum_{m=1}^{n_b} \lambda_{Q_m}(Q_m - Q_{G_m} + Q_{D_m})$$

$$+ \sum_{L=1}^{n_L} \mu_L(F_L - F_L^{max}) + \sum_{k=1}^{n_g} \mu_{G_k}^-(P_{G_k}^{min} - P_{G_k}) + \sum_{k=1}^{n_g} \mu_{G_k}^+(P_{G_k} - P_{G_k}^{max})$$

$$+ \sum_{k=1}^{n_g} \mu_{G_k}^-(Q_{G_k}^{min} - Q_{G_k}) + \sum_{k=1}^{n_g} \mu_{G_k}^+(Q_{G_k} - Q_{G_k}^{max}) + \sum_{m=1}^{n_b} \mu_{V_m}^-(V_m^{min} - V_m)$$

$$+ \sum_{m=1}^{n_b} \mu_{V_m}^+(V_m - V_m^{max}) \tag{5.10}$$

where λ and μ are Lagrangian multipliers vectors associated with equality constraints and inequality constraints respectively obtained by OPF solution. The OPF solution to equation (5.10) is found with the help of MATPOWER 5.0 used in MATLAB environment.

5.2.3 Congestion zones identification

In this chapter, the management of congestion is achieved through inter-zonal congestion alleviation. Congestion zones are nothing but a group of buses connected across a line. The transmission network bifurcation into different zones based on locational marginal price is proposed. LMP gives an economic signal to the electricity market and is therefore preferred these days in most of the electricity market to manage congestion. Thus the LMP provides a measure to identify different congestion zones. The different LMP based methods for congestion zones identification are explained below.

5.2.3.1 Average LMP method

One of the basic and easiest method to identify congestion zones using LMP is based on the average LMP [139]. The LMP calculated at each bus is approximately same for uncongested system. When congestion occurs, the LMP at each bus is different from each other. The busses having small variation in their LMP are segregated in a single zone and the average value of the LMP is calculated for that zone. Similarly other congestion zones are identified. The zone having highest value of average LMP is identified the most sensitive congestion zone.

5.2.3.2 LMP difference method

Another LMP based method to identify the congestion zones makes use of the difference of LMP of buses. Congestion zones for a given system are defined based on LMP difference across a line. Congestion zones are nothing but a group of buses connected across a line, selected based on LMP difference across that line given by equation (5.11).

$$\Delta LMP_L = LMP_m - LMP_n \quad ; \quad L = 1, 2, \ldots, n_L \tag{5.11}$$

where ΔLMP_L is the LMP difference across line L, LMP_m and LMP_n are the LMPs at bus-m and bus-n respectively.

The zone having high and non-uniform LMP difference between buses across a line has been identified as zone of type-1 and the zones having low and uniform LMP difference between buses across a line are defined as zone of type-2 and so on. Therefore, the transactions in the congestion zone-1 have critical and unequal impact on the LMP. The other congestion zones are farther from interested congested line. Hence, any transaction outside the most sensitive zone-1 will have little effect on line flow and LMP. Therefore, the identification of zones of congestion will lead to the reduction of computational burden involved in congestion management schemes required for the transmission loading relief which is described in next section.

5.2.4 Optimal allocation of distributed generation

The identification of congestion zones segregates the zone of interest having greater impact on line flow with more prone to congestion. The different zonal congestion management schemes have been adopted in [129]-[134] based on real and reactive power distribution factors. Although most of these methods utilize the generation side approaches for congestion alleviation, but in the new era of competitive electricity market, demand side approach are getting more attention as it alleviates congestion more effectively thereby improving the reliability and security of the power system [95]-[100]. Therefore distributed generations (DGs) are getting more interest in restructured power system as it can be generally located in load pockets and can be considered as negative power demands. DGs can quickly respond to the changing conditions of competitive electricity market and therefore has got an augmented interest for power system operation and planning in deregulated environment. Other benefits of DGs include easy installation and simple operation with low capital cost. Its strategical location and operation in system reduce losses, improve voltage profile, defer system upgrades and improve reliability of the system. With all these benefits DGs are being extensively used for congestion management in restructured power system [101]-[103].

However, the investment in DG as well as the electricity generation cost using it involves high cost, therefore its appropriate location and size play an important role to mitigate congestion optimally [104]-108]. It can directly affect the electricity market economics. The next section describes the different methods for optimal placement of DG based on LMP.

5.2.4.1 Highest LMP method

A simple method that can be employed to place DG optimally for congestion alleviation is highest LMP method [139]. Under normal operation of a power system, LMP at all buses remain almost same, but as congestion occurs, the LMPs at all buses will become different. In highest LMP method for optimal DG placement, buses are ranked on the basis of their LMPs such that the bus having highest LMP would be ranked 1 and so on.

5.2.4.2 LMP difference method

Although the highest LMP method of DG placement is simple, but it may give rise to a situation that the congestion increase in the network. Therefore, this method could not be reliably used for DG placement. Another method which is based on difference of LMPs of two buses across a line, known as " LMP difference method" can be utilized more efficiently and reliably to place a DG optimally in order to mitigate congestion. The optimal placement of TCSC using LMP difference method for congestion management has been discussed in [104]. Since LMP difference across a congested line will be highest as compared to other lines, therefore this method can more effectively reflect the list of potential locations for DG placement. Therefore, a priority list based on LMP difference across a line is formed to find the potential location for optimal DG placement. The LMP difference across a line is given by equation (5.12).

$$\Delta LMP_{kl} = LMP_k - LMP_l \quad ; \quad kl = 1,2,\ldots,n_L \quad (5.12)$$

where ΔLMP_{kl} is the LMP difference across line-kl, LMP_k and LMP_l are the LMPs at bus-k and bus-l respectively.

5.3 Results and discussions

The proposed LMP difference based method is used to identify the congestion zones as well as to optimally place the DGs. Initially, the solution space is reduced by considering a list of bus candidate location. Since the buses having more generation capacity than their demand have low LMP, therefore these are not considered as potential locations for DG placement. Hence buses are examined for potential location of DG according to equation (5.13) such that it has either no generating unit or it has low generating capacity as compared to its load.

$$P_{G_k} \le P_{D_k} \quad ; \quad k = 1,2,\ldots\ldots,n_b \quad (5.13)$$

The robustness of proposed methodology is analysed on IEEE 14-bus system and IEEE 57-bus system. Since a number of DG technologies with varying operating characteristics are nowadays available in the market, therefore assumptions are made for cost characteristics of DG in order to accommodate this variation [112]. The DG is considered to inject only real power of 5 MW.

5.3.1 Results for IEEE 14-bus system

Table 5.1 to Table 5.4 shows the results for IEEE 14-bus system.

Table 5.1 shows the LMP difference across different lines obtained from OPF solution. It shows that LMP difference across line-1 to line-5 is high and non-uniform as compared to other lines. Therefore these lines are more prone to congestion as compared to other lines. Hence line-1 to line-5 being the most congestion sensitive line, the buses connecting them are considered to be grouped in zone-1 (most congestion sensitive zone) while the buses connecting remaining lines are grouped in zone-2 as their LMP difference is low and uniform as shown in Table 5.2. Identification of zones based on LMP difference is also illustrated in Figure 5.1.

Table 5.1 LMP difference across lines for IEEE 14-bus system

Line No.	From Bus-To Bus	LMP Difference ($/MWh)	Line No.	From Bus-To Bus	LMP Difference ($/MWh)
1	1-2	0.6	11	6-11	0.3
2	1-5	1.3	12	6-12	0.4
3	2-3	0.6	13	6-13	0.5
4	2-4	0.9	14	7-8	0.0
5	2-5	0.7	15	7-9	0.0
6	3-4	0.3	16	9-10	0.2
7	4-5	0.2	17	9-14	0.6
8	4-7	0.0	18	10-11	0.1
9	4-9	0.0	19	12-13	0.1
10	5-6	0.0	20	13-14	0.4

Table 5.2 Congestion zones identification based on LMP difference for IEEE 14-bus system

Congestion Zones	Bus No.
Zone-1	1, 2, 3, 4, and 5
Zone-2	6, 7, 8, 9, 10, 11, 12, 13, and 14

A load flow analysis of IEEE 14-bus system gives that line-3 (connected between bus-2 and bus-3) are congested. The power flow in line-3 is 35.48 MW which is above its power transfer limit. This line lies in the most sensitive congestion zone as shown in Table 2.

After the identification of most sensitive congestion zone-1, the congestion is managed by optimally placing the DG on bus of that particular zone. The buses across lines in congestion zone-1, having high LMP difference, are the potential locations for DG placement with the condition that the generation at those buses is less than their demand, thereby satisfying equation (5.13). Table 5.1 and 1 Table 5.2 show that line-2, line-3 and line- 4 in zone-1 have high LMP difference across them and are therefore more prone to congestion. Therefore the busses connecting these lines are the potential locations for DG placement in order to alleviate congestion. But, since buse-1, bus-2, and bus-3 does not satisfy equation (5.13), therefore these buses cannot be considered for DG placement. Only bus-4 and bus-5 in most sensitive congestion zone-1 are considered for DG placement. We can allocate DG to these buses separately and find the best location for congestion management. But in larger system, it would be difficult and time taking process to separately place the DG at all potential bus locations and find the optimal location. Therefore a method for optimal location of DG based on LMP difference is adopted such that it could be found in no time. From the potential locations for DG placement, the line connecting bus-5 (line-2) has the highest LMP difference. Therefore DG is placed at bus-5 for congestion management and the results are shown in Table 5.3 which depicts that both the system generation cost as well as the system congestion cost decreases when DG is implemented in most sensitive congestion zone.

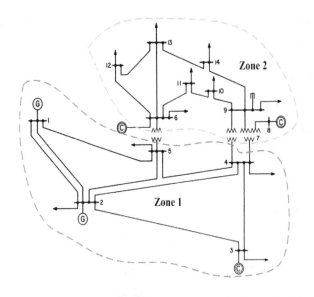

Figure 5.1 Congestion zones identification based on LMP difference for IEEE 14-bus system

Table 5.3 Results for IEEE 14-bus system

DG Location	System Generation Cost ($/hr)	System Congestion Cost ($/hr)
Without DG	3461.2	239.2
Bus-5	3221.0	198.1

After optimally placing DG in most sensitive congestion zone-1, the load flow analysis gives that the power flow on congested line-3 reduces to 34.98 MW which is well within its transfer limit. Thus the congestion is effectively alleviated with the optimal placement of DG in most sensitive congestion zone.

Also, to analyse the effectiveness of zonal based congestion management using LMP difference, the DG is placed to other potential location of congestion zone-1 i.e bus-4 as well as buses of congestion zone-2 which is considered less prone to congestion as compared to congestion zone-1 and the results are shown in Table 5.4. The minimum generation cost due to placement of DG at different buses in congestion zone-2 is obtained for bus-14. Therefore only bus-14 is considered here for analysing the effectiveness of proposed methodology.

Table 5.4 reveals that DG allocation at bus-5 in congestion zone-1 gives less generation cost as compared to when it is placed at bus-4 in same congestion zone. It also reveals that DG allocation in congestion zone-1 gives low system generation cost for both the potential locations of DG placement as compared to when it is placed at any bus locations of congestion zone-2. Hence the identification of congestion zones and placement of DG based on LMP difference provides management of congestion in a better and easier manner and is illustrated in Figure 5.2.

Table 5.4 Generation cost for IEEE 14-bus system in different zones

Congestion Zones	DG Location	System Generation Cost ($/hr)
Zone-1	Bus-5	3221.0
Zone-1	Bus-4	3221.7
Zone-2	Bus-14	3336.6

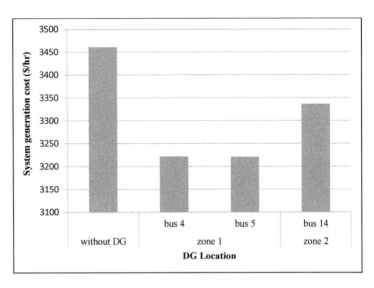

Figure 5.2 Generation cost for IEEE 14-bus system

5.3.2 Results for IEEE 57-bus system

.Table 5.5 shows the LMP difference obtained across each line of IEEE 57-bus system on the basis of which congestion zones are identified. The LMP difference across lines connecting bus-17 is high. Therefore the lines connecting nearby buses have also high and non-uniform LMP. Hence bus-17 and nearby buses are grouped in congestion zone-1 in which the lines connecting these buses are most sensitive to congestion and is shown in Table 5.6.

The other zones have buses connecting the lines of less LMP difference across them. Zone-2 has also buses connecting lines of high LMP difference, but these lines have more uniform LMP difference. Therefore zone-2 is considered to be less sensitive to congestion. The remaining zones have buses connected to line with low and uniform LMP difference across them and hence are less sensitive to congestion as compared to congestion zone-1 and zone-2. An illustration of the different congestion zone is shown in Figure 5.3.

Table 5.5 LMP difference across lines for IEEE 57-bus system

Line No.	From Bus-To Bus	LMP Difference ($/MWh)	Line No.	From Bus-To Bus	LMP Difference ($/MWh)	Line No.	From Bus-To Bus	LMP Difference ($/MWh)
1	1-2	0.55	28	14-15	1.08	55	41-42	2.63
2	2-3	3.04	29	18-19	2.32	56	41-43	0.30
3	3-4	1.16	30	19-20	0.56	57	38-44	0.75
4	4-5	1.35	31	21-20	0.20	58	15-45	0.46
5	4-6	1.63	32	21-22	0.03	59	14-46	0.21
6	6-7	8.09	33	22-23	0.27	60	46-47	1.11
7	6-8	8.79	34	23-24	3.50	61	47-48	0.33
8	8-9	4.36	35	24-25	0.39	62	48-49	0.41
9	9-10	3.26	36	24-25	0.39	63	49-50	0.59
10	9-11	2.00	37	24-26	0.64	64	50-51	1.62
11	9-12	4.83	38	26-27	1.29	65	10-51	0.09
12	9-13	3.10	39	27-28	0.14	66	13-49	1.09
13	13-14	0.15	40	28-29	0.35	67	29-52	0.16
14	13-15	1.23	41	7-29	1.42	68	52-53	0.30
15	1-15	3.83	42	25-30	1.03	69	53-54	4.78
16	1-16	5.37	43	30-31	1.22	70	54-55	4.98
17	1-17	21.58	44	31-32	1.87	71	11-43	0.10
18	3-15	0.25	45	32-33	0.18	72	44-45	2.28
19	4-18	0.06	46	33-32	1.35	73	40-56	0.90
20	4-18	0.06	47	32-35	0.52	74	56-41	3.78
21	5-6	0.29	48	35-36	0.70	75	56-42	1.15
22	7-8	16.88	49	36-37	0.48	76	39-57	0.27
23	10-12	1.57	50	37-38	1.24	77	57-56	0.75
24	11-13	1.10	51	37-39	0.07	78	38-49	1.18
25	12-13	1.73	52	36-40	0.01	79	38-48	0.77
26	12-16	1.42	53	22-38	0.53	80	9-55	1.19
27	12-17	14.80	54	11-41	0.41			

Table 5.6 Congestion zones identification based on LMP difference for IEEE 57-bus system

Congestion Zones	Bus No.
Zone-1	1, 12, 13, 14, 15, 16, 17, 44, and 45
Zone-2	6, 7, 8, 9, 52, 53, 54, and 55
Zone-3	2, 3, 4, 5, 18, 19, 20, 21, 22, 23, 24, 25, 26, 27, 28, and 29
Zone-4	10, 11, 30, 31, 32, 33, 34, 35, 36, 37, 38, 39, 40, 41, 42, 43, 46, 47, 48, 49, 50, 51, 56, and 57

A load flow analysis of IEEE 57-bus system gives that the line-17 (connected between bus-1 and bus-17) is congested. The power flow on line-17 is 79.19 MW which is above its thermal limit.

After the identification of congestion zones, the optimal location for DG placement is identified based on LMP difference in congestion zone-1. Since bus-1 and bus-12 do not satisfy equation (5.13), therefore remaining buses in congestion zone-1 are potential locations for DG placement. But the lines connecting bus 1 and 17 has highest LMP difference, and also bus is among the potential locations. Therefore, bus-17 is selected as an optimal location for DG placement in order to manage congestion and the results are given in Table 5.7 which shows that both the system generation cost as well as the system congestion cost decreases significantly when DG is implemented in most sensitive congestion.

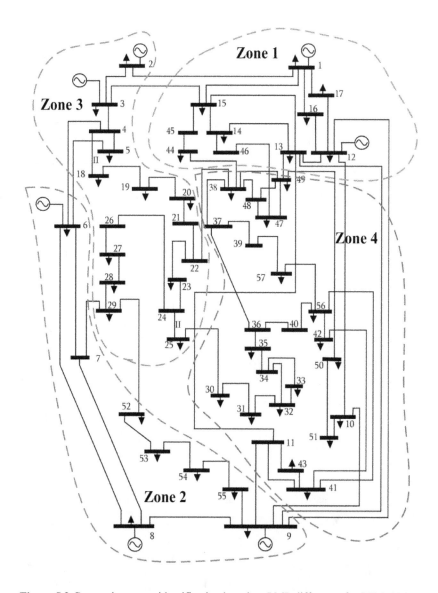

Figure 5.3 Congestion zones identification based on LMP difference for IEEE 57-bus system

Table 5.7 Results for IEEE 57-bus system

DG Location	System Generation Cost ($/hr)	System Congestion Cost ($/hr)
Without DG	41920.8	4610.1
Bus-17	41638.6	3584.4

After optimally placing DG in most sensitive congestion zone-1, the load flow analysis gives that the power flow on congested line-17 reduces 39.23 MW which is well within its thermal limit. Thus the congestion is effectively alleviated with the optimal placement of DG.

Also to analyze the effectiveness of proposed methodology in larger systems, the DG is placed in other congestion zones and the results for its location of minimum system generation cost in a particular zone are shown in Table 5.8.

Table 5.8 Generation cost for IEEE 57-bus system in different zones

Congestion Zones	DG Location	System Generation Cost ($/hr)
Zone-1	Bus-17	41638.6
Zone-2	Bus-7	41666.0
Zone-3	Bus-23	41670.9
Zone-4	Bus-56	41671.9

Table 5.8 reveals that the placement of DG in most sensitive congestion zone-1 gives less system generation cost as compared to when it is placed in other congestion zones. Hence the allocation of DG in most sensitive congestion zone-1 with the proposed methodology manages congestion more effectively and efficiently as compared to its allocation in other congestion zones as illustrated in Figure 5.4.

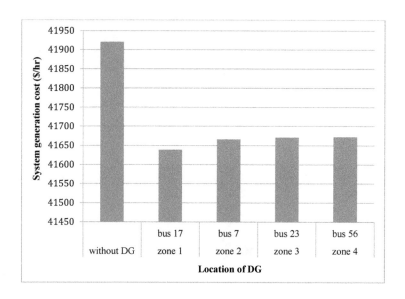

Figure 5.4 Generation cost for IEEE 57-bus system

5.4 Conclusion

A new approach to zonal congestion management in pool market is proposed. The identification of different congestion zones is based on LMP difference between buses connecting a line. The most sensitive congestion zone is one which groups the buses connecting lines of high and non-uniform LMP difference across them. The congestion is managed by optimally allocating the DG in most sensitive congestion zone. The optimal allocation of DG is also based on LMP difference. To analyze the effectiveness of zonal based congestion management, the DG is also allocated to other zones which are considered less sensitive to congestion. The robustness of proposed methodology is tested on IEEE 14-bus system and IEEE 57-bus system and it is found to be efficient for both small as well as large power systems.

CHAPTER – 6

CONGESTION MANAGEMENT USING FACTS DEVICE

6.1 Introduction

With the deregulation of power system and problems in securing new rights of way have driven the electricity sector to adopt new and major technological developments based on high voltage, high current solid state controllers. This results in the development of Flexible AC Transmission System (FACTS) devices which regulate the power flow by altering the transmission line parameters such as X (line reactance), V (voltage magnitude) and δ (voltage angle) [57]. These devices can also be used for voltage stability improvement, transient stability improvement, sub-synchronous resonance mitigation etc [57]-[58]. These devices reduce congestion by increasing power flow in the existing lines [59]-[60].

According to IEEE, FACTS is defined as "as a power electronic-based system and other static equipment that has the ability to enhance controllability, increase power transfer capability".

The basic concept of FACTS devices has been introduced by Narain G. Hingorani in 1988 [140] and since then it has been widely adopted around the world for various power system applications. In the current scenario of deregulated electricity market, electricity generating utilities and system operators around the world has a challenge of increased demands for bulk power transmission, low cost power delivery and higher reliability. Such issues can be efficiently taken care by FACTS devices to some extent and that's why it has been extensively used round the globe. The versatility in its connection with power system, i.e in series, shunt or in combination of both, provides appropriate and required compensation to the power systems. The power regulating feature of FACTS devices has made it popular to utilize it for managing congestion in deregulated environment of power system.

The following sections in this chapter deal with theory, modelling and use of FACTS devices for congestions mitigation.

6.2 Basic concepts of FACTS

6.2.1 An overview

The active power transmitted between systems is given by [141]:

$$P = \frac{V_1 V_2}{X} \sin(\delta_1 - \delta_2) \qquad (6.1)$$

where V_1 and V_2 are voltages at two ends of transmission line, X is the equivalent reactance of the transmission line and (δ_1 - δ_2) is the phase angle difference between two voltages. From equation (6.1) it is evident that the power transmitted is influenced by three parameters: voltage, reactance and angle difference. FACTS devices can influence one or more of these parameters thereby controlling power flow. The control of first two parameters can be used to increase power limits while last parameter can be used to control power flow in loops. This is shown in Figure 6.1.

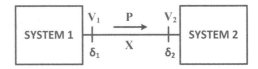

Figure 6.1 Power flow between two systems

Figure 6.2 shows various FACTS devices used for controlling one or more of these three parameters. For example, Static Var Compensator (SVC) acts as controllable shunt susceptance (capacitive or inductive) and influences the magnitude of voltage by generating or consuming reactive power [142]-[143].

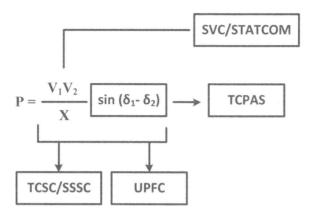

Figure 6.2 Functions of different FACTS devices

Thyristor Controlled Phase Angle Shifter (TCPAS) changes the phase angle difference between two ends of transmission line while Thyristor Controlled Series Compensator (TCSC) modulates reactance of transmission line. Unified Power Flow Controller (UPFC) controls all the three parameters simultaneously. These devices can be used to control power flow, voltage magnitude and other system parameters [143]-[145].

If a VSC is connected to a transmission system in shunt via a transformer, it can generate or absorb reactive power from the bus to which it is connected such a controller is called Synchronous Static Compensator (STATCOM) and is used for voltage control in transmission systems. The major advantage of a STATCOM as compared to a SVC is its reduced size and cost due to elimination of bulky capacitor and reactor banks.

If a VSC is used as a series device by connecting it to a transmission line through a series transformer, it is known as Static Synchronous Series Compensator or SSSC. This controller can also generate or absorb reactive power from the line to which it is connected and hence change the series impedance of line. SSSC can be assumed to be a continuously variable series capacitor or inductor and hence, can be used to control power flow in line. A UPFC is a combination of STATCOM and SSSC.

6.2.1 Classification of FACTS devices

Table 6.1 shows list of FACTS devices which can be classified into following three classes [140]-[141]:

1. Shunt connected devices controlling voltage.
2. Series connected devices controlling power flow.
3. Hybrid devices controlling both voltage and power flow

Table 6.1 Classification of FACTS devices [140]-[141]

Type of Device	Name of Device	Main Function	Switching Element	System Parameter
Shunt Connected	SVC	Voltage	Thyristor	Variable susceptance
	STATCOM	Voltage control	GTO	Variable voltage
Series Connected	TCSC	Power flow control	Thyristor	Variable reactance
	SSSC	Power flow control	GTO	Variable voltage
Hybrid	UPFC (series-shunt)	Voltage and power flow control	GTO	Variable voltage
	IPFC (series-series)	Power flow control	GTO	Variable voltage
	TCPAS (series-shunt)	Power flow control	Thyristor	Phase control (series quadrature voltage injection)

TCSC is one of the widely used FACTS devices around the world. Its simple construction and implementation as well as low cost as compared to other FACTS devices make it preferable over others for congestion management [146]-[148]. Therefore in this research work also, TCSC is considered for the purpose of managing congestion which has been discussed in the next section.

6.3 Thyristor Controlled Series Capacitor (TCSC): A series FACTS device

Power flow in a transmission network can be adjusted by varying the net series reactance. Application of series capacitor to increase transmission line capacity is a well-known method of series compensation which helps to reduce net series reactance thereby allowing flow of additional power through the lines. However, the conventional methods of series compensation use capacitors with mechanical switches such as circuit breakers over a limited range while compensation using thyristor controllers rapidly controls the line compensation over a continuous range with flexibility.

A Thyristor Controlled Series Capacitor (TCSC) [148] is shown in Figure 6.3. It consists of Thyristor Controlled Reactor (TCR) in parallel with capacitor segments of a series capacitor bank which allows the capacitive reactance to be smoothly controlled over a wide range and switched upon command to a condition where bi-directional thyristor pairs conduct continuously and insert an inductive reactance into the line.

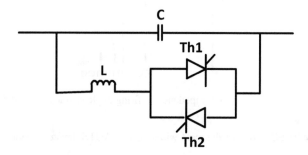

Figure 6.3 Thyristor Controlled Series Capacitor (TCSC)

Figure 6.4 shows impedance characteristic of TCSC, which illustrates that both capacitive as well as inductive region are possible through varying firing angle (α)

[148]. For firing angle α such that $90^0 < α < α_{Llim}$, the region is inductive. Similarly for firing angle α between $α_{Clim}$ and 180^0 (i.e. $α_{Clim} < α < 180^0$) the region is capacitive. For firing angle between $α_{Llim}$ and $α_{Clim}$ (i.e. $α_{Llim} < α < α_{Clim}$), there exists a resonance region. While selecting inductance, X_L should be sufficiently smaller than that of the capacitive reactance X_C. If X_C is smaller than the X_L, then only capacitive region is possible in impedance characteristics. In any shunt network, the effective value of reactance follows the lesser reactance present in the branch. So only one capacitive reactance region will appear. Also X_L should not be equal to X_C value; or else a condition of resonance develops that result in infinite impedance which is an unacceptable condition.

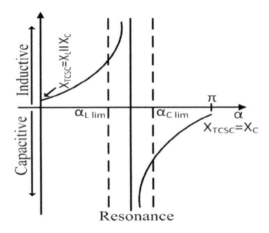

Figure 6.4 Impedance vs firing angle characteristic of TCSC

The TCSC is one of the most widely used FACTS device. Besides the application of TCSC for increasing power transfer capability of transmission line and reducing system losses, there are various other applications which can be can be summarised as follows [149]-[152].

- Damping of the power swings from local and inter area oscillations.
- Suppression of sub-synchronous oscillations.
- Voltage support.
- Reduction of short circuit current.

6.3.1 Modelling and implementation of TCSC

Figure 6.5 shows a π-equivalent transmission line model connected between bus-i and bus-j.

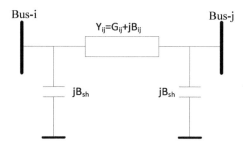

Figure 6.5 Transmission line model

If $V_i \angle \delta_i$ and $V_j \angle \delta_j$ are the voltages at bus-i and bus-j respectively, the equations for active and reactive power flow from bus-i to bus-j can be given by equation (6.2) and (6.3) respectively.

$$P_{ij} = V_i^2 G_{ij} - V_i V_j [G_{ij} \cos(\delta_{ij}) + B_{ij} \sin(\delta_{ij})] \tag{6.2}$$

$$Q_{ij} = -V_i^2 (B_{ij} + B_{sh}) - V_i V_j [G_{ij} \sin(\delta_{ij}) + B_{ij} \cos(\delta_{ij})] \tag{6.3}$$

where, $\delta_{ij} = \delta_i - \delta_j$. Similarly, the power flows from bus-j to bus-i are given by equations (6.4) and (6.5).

$$P_{ji} = V_j^2 G_{ij} - V_i V_j [G_{ij} \cos(\delta_{ij}) + B_{ij} \sin(\delta_{ij})] \tag{6.4}$$

$$Q_{ji} = -V_j^2 (B_{ij} + B_{sh}) - V_i V_j [G_{ij} \sin(\delta_{ij}) + B_{ij} \cos(\delta_{ij})] \tag{6.5}$$

The application of TCSC in a transmission network can be visualized as a control reactance connected in series to the specific transmission line. Figure 6.6 shows the transmission network model with a TCSC. During steady-state condition, a TCSC can be taken as a static capacitor/reactor with impedance jX_C [75]-[77].

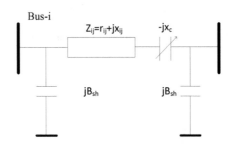

Figure 6.6 Transmission line model with TCSC

With the implementation of TCSC, the power flow from bus-i to bus-j is modified as given in equation (6.6) and (6.7).

$$P'_{ij} = V_i^2 G'_{ij} - V_i V_j [G'_{ij} \cos(\delta_{ij}) + B'_{ij} \sin(\delta_{ij})] \tag{6.6}$$

$$Q'_{ij} = -V_i^2 (B'_{ij} + B'_{sh}) - V_i V_j [G'_{ij} \sin(\delta_{ij}) + B'_{ij} \cos(\delta_{ij})] \tag{6.7}$$

where,

$$G'_{ij} = \frac{r_{ij}}{r_{ij}^2 + (x_{ij} - x_c)^2}$$

and

$$B'_{ij} = \frac{-(x_{ij} - x_c)}{r_{ij}^2 + (x_{ij} - x_c)^2}$$

Generally, a congestion management problem employs static model of FACTS device injecting power at sending and receiving end of line [77]-[80]. According to this model FACTS device can be represented as PQ element injecting definite amount of power to the specific node. Figure 6.7 shows the power injection model of TCSC.

The real power injected at bus-i and bus-j due to implementing TCSC is given by equation (6.8) and (6.9).

Figure 6.7 Power injection model

$$P_i' = V_i^2 \Delta G_{ij} - V_i V_j [\Delta G_{ij} \cos\delta_{ij} + \Delta B_{ij} \sin\delta_{ij}] \quad (6.8)$$

$$P_j' = V_j^2 \Delta G_{ij} - V_i V_j [\Delta G_{ij} \cos\delta_{ij} + \Delta B_{ij} \sin\delta_{ij}] \quad (6.9)$$

Similarly the reactive power injected at bus-i and bus-j after implementing TCSC is given by equation (6.10) and (6.11).

$$Q_i' = -V_i^2 \Delta B_{ij} - V_i V_j [\Delta G_{ij} \sin\delta_{ij} + \Delta B_{ij} \cos\delta_{ij}] \quad (6.10)$$

$$Q_j' = -V_j^2 \Delta B_{ij} - V_i V_j [\Delta G_{ij} \sin\delta_{ij} + \Delta B_{ij} \cos\delta_{ij}] \quad (6.11)$$

Where,

$$\Delta G_{ij} = \frac{x_c r_{ij}(x_c - 2x_{ij})}{(r_{ij}^2 + x_{ij}^2)(r_{ij}^2 + (x_{ij} - x_c)^2)}$$

and

$$\Delta B_{ij} = \frac{-x_c(r_{ij}^2 - x_{ij}^2 + x_c x_{ij})}{(r_{ij}^2 + x_{ij}^2)(r_{ij}^2 + (x_{ij} - x_c)^2)}$$

6.4 Problem formulation for congestion management using TCSC

The main objective of this work is to manage congestion by optimally placing TCSC in the power system network which is achieved by minimizing the cost of installation of FACTS device [68] -[87] along with penalty for violation of line flow limits and bus voltage limits as shown in equation (6.12).

$$\text{Min } [C_i(P_i) + C_{TCSC} + \lambda_1.VLV + \lambda_2.FLV] \tag{6.12}$$

where

$$C_{TCSC} = C_t \times S \times 1000 \quad (\$/hr) \tag{6.13}$$

$$C_t = 0.0015\, S^2 + 0.713 S + 153.75 \quad (\$/KVAR) \tag{6.14}$$

$$S = |Q_1 - Q_2| \tag{6.15}$$

where, C_t is the unit cost of TCSC

S is the operating range of TCSC in MVAR

Q1 and Q2 are the reactive power flow in the line before and after installation of TCSC

P_L^2 is the power flow in line k to which TCSC is connected

λ_1 and λ_2 are the penalty coefficients in the range of 10^5 to 10^8

VLV is the voltage violation factor

FLV is the line flow limit violation factor

The objective function comprises of two parts. The first part is the installation cost of TCSC whereas the second part is composed of penalty cost due to violation of bus voltage limit and the line flow limit which are given as:

$$VLV = \left(\frac{V_b - V_{ref}}{V_{ref}}\right)^2, \quad if\ V_b < V_{ref}^{min}\ or\ V_b > V_{ref}^{max} \tag{6.16}$$

$$= 0 \quad if\ V_{ref}^{min} < V_b < V_{ref}^{max} \tag{6.17}$$

$$FLV = \left(\frac{P_{ij} - P_{ij}^{max}}{P_{ij}^{max}}\right)^2, \quad if\ P_{ij} > P_{ij}^{max} \tag{6.18}$$

$$= 0 \quad if\ P_{ij} < P_{ij}^{max} \tag{6.19}$$

The reactance of TCSC is chosen such that $X_{ck}^{min} < X_{ck} < X_{ck}^{max}$. For static model of TCSC, the maximum compensation allowed is 70% of the reactance of line [77]. The voltages V_{ref}^{min} is taken as 0.94 pu while V_{ref}^{max} is taken as 1.06 pu.0

6.5 Optimal placement of TCSC

TCSC involves a heavy investment for its installation. Therefore, its appropriate location and size play a very vital role in managing congestion efficiently. Otherwise, it would not be proved beneficial as compared to other methods of congestion management. Therefore, the size and location of TCSC must be chosen with utmost care [68]-[87].

In this research work, the TCSC is optimally placed in the system considering the line flow sensitivity factor which is defined as a change in real power flow in a transmission line connected between bus-i and bus-j due to change in control parameter of TCSC.

Since the real power flow of a transmission line changes with the change in its reactance, the real power flow in the network paths changes due to the change in series reactance of the line by placing TCSC. This change in real power flow is a function of control parameter (i.e. reactance setting) of TCSC. Thus change in real power flow of a line due to change in control parameter of TCSC gives an indication for optimal placement of TCSC in managing congestion.

Mathematically, the line flow sensitivity factor with respect to the parameters of TCSC placed at line-k can be defined as:

$$LSF_c^k = \frac{\partial P_{LT}}{\partial X_{ck}}\bigg|_{X_{ck}=0} \qquad (6.20)$$

where, P_{LT} is the real power flow in line connected between bus i and bus j

X_{ck} is the control parameter of TCSC

The lines with high negative values of line flow sensitivity factor are the potential locations for placing TCSC in the network in order to manage congestion efficiently.

Equation (6.20) is calculated by differentiating the power flow in a line with respect to TCSC control parameters which is given as:

$$LSF_c^k = \frac{\partial P_{LT}}{\partial X_{ck}}\bigg|_{X_{ck}=0}$$

$$= V_i^2 \left(\frac{-2 r_{ij} x_{ij}}{(r_{ij}^2 + x_{ij}^2)^2}\right) - V_i V_j \left[\left(\frac{-2 r_{ij} x_{ij}}{(r_{ij}^2 + x_{ij}^2)^2} \cos \delta_{ij}\right) - \left(\frac{r_{ij}^2 - x_{ij}^2}{(r_{ij}^2 + x_{ij}^2)^2} \sin \delta_{ij}\right)\right]$$

(6.21)

$$= C_{ij}\left[-V_i^2 + V_j V \cos \delta_{ijj}\right] - D_{ij}\left[V_i V_j \sin \delta_{ij}\right] \quad (6.22)$$

where

$$C_{ij} = \left(\frac{-2 r_{ij} x_{ij}}{(r_{ij}^2 + x_{ij}^2)^2}\right)$$

$$D_{ij} = \left(\frac{r_{ij}^2 - x_{ij}^2}{(r_{ij}^2 + x_{ij}^2)^2}\right)$$

Once the optical location of TCSC is found, its optimal setting for congestion alleviation is found using PSO algorithm and can be represented by the flow chart given in Figure 6.8.

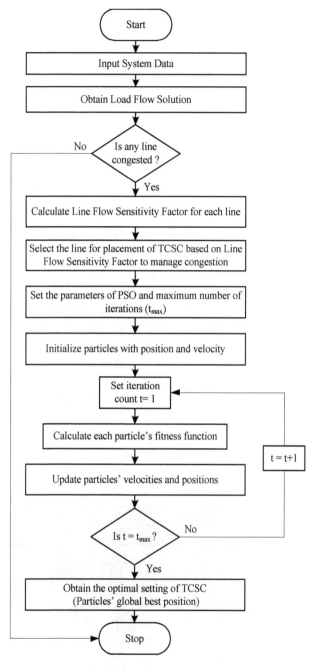

Figure 6.8 Flow chart for optimal setting of TCSC

6.6 Results and discussions

The FACTS device should be placed on the most sensitive line. With the sensitive factor computed for TCSC, the TCSC should be placed in a line having the most negative line flow sensitivity factor. The proposed method for congestion management using TCSC is implemented and tested with IEEE 14-bus system in order to analyse its effectiveness and robustness. Optimizations are carried out with PSO developed in MATLAB language. The value of various parameters taken for PSO are same as taken in chapter 1 (i.e. $w_{min} = 0.4$, $w_{max} = 0.9$, $c_1=c_2=2.0$). Maximum iterations is set to 500 and particle size is taken as 70.

Table 6.2 Power flow for IEEE 14-bus system without TCSC

Line No.	From Bus	To Bus	P (MW)	P_{max} (MW)
1	1	2	92.38	120
2	1	5	51.39	65
3	2	3	**35.48**	35
4	2	4	40.19	65
5	2	5	33.52	50
6	3	4	-0.11	65
7	4	5	-29.09	45
8	4	7	30.76	55
9	4	9	17.56	32
10	5	6	46.20	55
11	6	11	8.90	18
12	6	12	8.01	32
13	6	13	18.08	32
14	7	8	0.00	32
15	7	9	30.76	35
16	9	10	9.85	32
17	9	14	8.95	32
18	10	11	-5.18	12
19	12	13	1.83	12
20	13	14	6.16	12

Table 6.2 shows the power flow for IEEE 14-bus system in base case which gives that power flow in line-3 connected between bus-2 and bus-3 is above its limit. The line flow sensitivity factor with respect to TCSC control parameter has been computed for congested line-3 and is shown Table 6.3.

From Table 6.3, it is observed that line-1, line-4 and line-7 (connected between bus-1 and bus-2, bus 2 and bus-4, bus-4 and bus-5 respectively) have high negative value of line flow sensitivity factor as compared to other lines. Therefore these lines have high priority for placement of TCSC in order to mitigate congestion. Line-7 having the highest negative value of sensitivity factor is selected for placement of TCSC to manage congestion. Figure 6.9 illustrates these potential locations for TCSC placement which have high negative values of line flow sensitivity factor.

Table 6.3 Line flow sensitivity factor of IEEE 14-bus system for congested line-3

Line No.	From Bus	To Bus	Sensitivity Factor	Line No.	From Bus	To Bus	Sensitivity Factor0
1	1	2	**-0.2642**	11	6	11	0.0038
2	1	5	-0.0094	12	6	12	0.0027
3	2	3	-0.0311	13	6	13	0.0078
4	2	4	**-0.0474**	14	7	8	0.0000
5	2	5	-0.0339	15	7	9	0.0864
6	3	4	0.0033	16	9	10	0.0318
7	4	5	**-0.3519**	17	9	14	0.0004
8	4	7	-0.0059	18	10	11	-0.0031
9	4	9	-0.003	19	12	13	-0.0001
10	5	6	-0.0094	20	13	14	0.0013

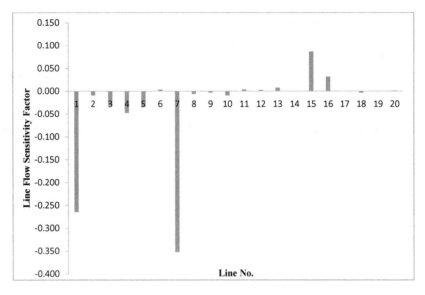

Figure 6.9 Line Flow Sensitivity Factor for IEEE 14-bus system

From the optimization algorithm, it has been found that for optimal setting of TCSC at 0.01582 pu, the congestion from the system is managed efficiently. For this setting of TCSC, the power flow in lines is shown in Table 6.4 which reveals that the power flow in line-3 which was earlier congested is within its limit after placing TCSC.

To analyse the effectiveness of the proposed sensitivity based approach for congestion management, TCSC is also placed in other potential locations i.e line-1 and line-4. The power flow for connecting TCSC in line-1 is shown in Table 6.5. The value of control parameter of TCSC is found as 0.01690 pu for which the congestion in line-3 is relieved as shown in Table 6.5.

Similarly, the power flow for placement of TCSC in line-4 is shown in Table 6.6, from which it can be observed that the congestion in line-3 is relieved with this location of TCSC. The value of control parameter of TCSC for this location is computed from optimization algorithm as 0.01623 pu.

Table 6.4 Power flow for IEEE 14-bus system with TCSC in line-7

Line No.	From Bus	To Bus	P (MW)	P$_{max}$ (MW)
1	1	2	91.73	120
2	1	5	52.03	65
3	**2**	**3**	**34.98**	**35**
4	2	4	39.11	65
5	2	5	34.49	50
6	3	4	-0.60	65
7	4	5	-31.15	45
8	4	7	31.07	55
9	4	9	17.75	32
10	5	6	45.67	55
11	6	11	8.58	18
12	6	12	7.98	32
13	6	13	17.92	32
14	7	8	0.00	32
15	7	9	31.07	35
16	9	10	10.17	32
17	9	14	9.15	32
18	10	11	-4.86	12
19	12	13	1.80	12
20	13	14	5.95	12

Table 6.5 Power flow for IEEE 14-bus system with TCSC in line-1

Line No.	From Bus	To Bus	P (MW)	P_{max} (MW)
1	1	2	88.57	120
2	1	5	55.13	65
3	2	3	**34.87**	35
4	2	4	38.89	65
5	2	5	31.77	50
6	3	4	-0.71	65
7	4	5	-30.83	45
8	4	7	30.67	55
9	4	9	17.52	32
10	5	6	46.31	55
11	6	11	8.97	18
12	6	12	8.02	32
13	6	13	18.12	32
14	7	8	0.00	32
15	7	9	30.67	35
16	9	10	9.78	32
17	9	14	8.90	32
18	10	11	-5.25	12
19	12	13	1.84	12
20	13	14	6.20	12

Table 6.6 Power flow for IEEE 14-bus system with TCSC in line-4

Line No.	From Bus	To Bus	P (MW)	P_{max} (MW)
1	1	2	93.04	120
2	1	5	50.75	65
3	**2**	**3**	**34.85**	**35**
4	2	4	42.47	65
5	2	5	32.53	50
6	3	4	-0.73	65
7	4	5	-27.63	45
8	4	7	30.80	55
9	4	9	17.60	32
10	5	6	46.10	55
11	6	11	8.84	18
12	6	12	8.01	32
13	6	13	18.05	32
14	7	8	0.00	32
15	7	9	30.80	35
16	9	10	9.91	32
17	9	14	8.98	32
18	10	11	-5.12	12
19	12	13	1.83	12
20	13	14	6.12	12

It is observed from Table 6.4, Table 6.5 and Table 6.6 that the placement of TCSC based on line flow sensitivity factor manages the congestion efficiently of congested line-3 which is also shown graphically in Figure 6.10, Figure 6.11 and Figure 6.12.

Figure 6.9 gives the graphical representation of power flow in different lines after placing TCSC in line-7 which demonstrates that congestion has been managed successfully accommodating all constraints for transmission of power. Similarly, Figure 6.10 and Figure 6.11 represent power flow in lines with placement of TCSC in line-1 and line-4 respectively.

Although the placement of TCSC at all potential locations manages the congestion successfully, it is necessary to perform the cost-benefit analysis for FACTS device in order to analyse its cost effective location among various potential locations. Therefore, installation cost of TCSC and hence the system generation cost is calculated using equation 6.12.

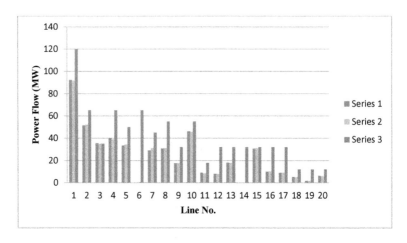

Figure 6.10 Power flow for IEEE 14-bus system with TCSC in line-7

Series 1: Power flow in lines without TCSC

Series 2: Power flow in lines with TCSC in line-7

Series 3: Maximum power flow limit of lines

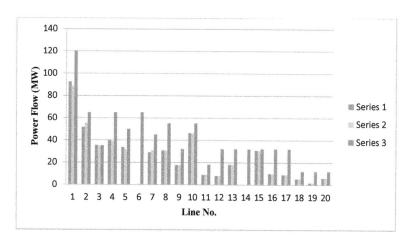

Figure 6.11 Power flow for IEEE 14-bus system with TCSC in line-1

Series 1: Power flow in lines without TCSC

Series 2: Power flow in lines with TCSC in line-1

Series 3: Maximum power flow limit of lines

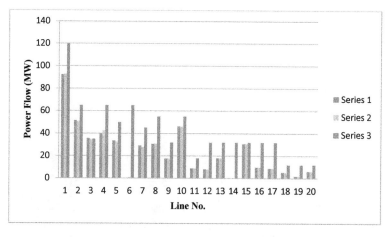

Figure 6.12 Line flow for IEEE 14-bus system with TCSC in line-4

Series 1: Power flow in lines without TCSC

Series 2: Power flow in lines with TCSC in line-4

Series 3: Maximum power flow limit of lines

Due to high cost involved with installation of FACTS devices, its investment cost should be analysed such that its placement gives the low installation cost. The installation cost of TCSC for different potential locations is calculated using equation 6.13-6.15 and is shown in Table 6.7 which reveals that placement of TCSC in line-7 is more economical as compared to other potential locations for TCSC placement and is graphically represented in Figure 6.13.

Table 6.7 Total Generation cost for IEEE 14-bus system

S.No.	Location of TCSC	Installation cost of TCSC ($/KVAR)
1	TCSC in line- 7	227.1
2	TCSC in line-1	246.2
3	TCSC in line-4	251.4

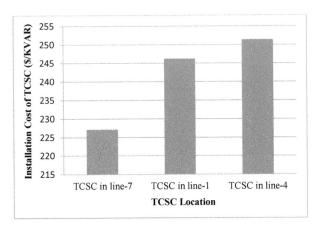

Figure 6.13 Installation cost of TCSC

The total generation cost without TCSC as well as with placement of TCSC at different potential locations is shown in Table 6.8.

Table 6.8 Total Generation cost for IEEE 14-bus system

S. No.	Location of TCSC	Total Generation Cost ($/hr)
1	Without TCSC	3782.8
2	TCSC in line- 7	3688.2
3	TCSC in line-1	3706.4
4	TCSC in line-4	3712.9

It is observed from Table 6.8 that the placement of TCSC in line-7 gives the minimum total generation cost as compared to its placement in other lines. Hence, TCSC placement in line-7 is more economical as compared to other locations which are also shown in Figure 6.14.

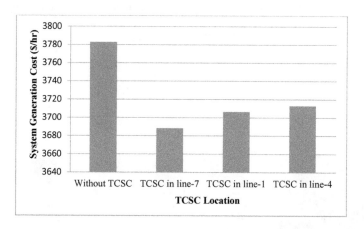

Figure 6.14 Total generation cost for IEEE 14-bus system

It is observed from the calculated installation cost of TCSC and total generation cost for different potential locations of TCSC placement, the placement of TCSC in line-7 is more economical as compared to other locations of TCSC placement.

6.7 Conclusion

In this chapter, impact of TCSC is analysed in managing congestion of transmission lines. A static model of TCSC is considered in problem formulation. Since FACTS devices involve a huge investment, therefor its optimal location plays an important role in order to achieve the market economics. A sensitivity based approach using power flow in lines with respect to TCSC control parameters is proposed to find the optimal location of TCSC. The line flow sensitivity factor is calculated for each line based on which the optimal location of TCSC has been determined. The effectiveness of the proposed method has also been analysed for minimizing the cost of TCSC installation as well as total generation cost. The proposed method has been tested on IEEE 14-bus system. It has been observed from test results that the placement of TCSC in most sensitive line determined from sensitivity analysis gives minimum installation cost of TCSC as well as minimum generation cost as compared to other potential locations for TCSC placement.

CHAPTER – 7

PARTICLE SWARM OPTIMIZATION WITH IMPROVED TIME VARYING ACCELERATION COEFFICIENTS FOR CONGESTION MANAGEMENT

7.1 Introduction

Supply of electricity to consumers at competitive rates is one of the main goals of deregulation of electricity market. However, presence of congestion in transmission lines hampers the accomplishment of this objective due to addition of congestion cost to the electricity tariff and also endangers system reliability. This issue has been addressed efficiently in previous chapters using three different techniques. However the efficient application of the previously discussed techniques depends upon the efficiency of the optimization algorithm used to solve the problem of congestion management. There are several optimization techniques proposed in literature which deal with solution of congestion management problem. These methods could be divided into two broad groups viz. classical optimization techniques and heuristic and meta heuristic techniques [82]-[83].

Classical optimization techniques such as Linear Programming (LP), Mixed Integer Linear Programming (MILP), Mixed Integer Non-Linear programming (MINLP), Quadratic Programming (QP), Non-Linear Programming (NLP) etc. use mathematical equations and iterative process to optimize the function which are good for unimodal and continuous functions. However, these techniques are time consuming, inefficient and solution tends to converge locally if the search space is large [84]. Other methods which are often used to optimize the function are Eigen-Values Analysis, Model

Analysis, Residue Method, Index Method, Lambda Iteration Method, Lagrange Multiplier Method, Projective Control Method etc. [85]-[87].

However with the development of artificial intelligence, heuristic and meta-heuristic methods are now widely used as an optimization algorithm. Genetic Algorithm (GA), Tabu Search (TS), Simulated Annealing (SA), Evolutionary Programming (EP), Differential Evolution (DE), Artificial Neural Network (ANN), Fuzzy Logic (FL), Ant Colony Optimization (ACO), Particle Swarm Optimization (PSO), Bacteria Swarming Optimization (BSO), Gravitational Search Algorithm (GSA), Harmony Search (HS) [88]-[92] etc. are some of the examples of this method. These stochastic algorithms are developed due to inability of deterministic algorithms such as Newton's method, Least Square Method, Linear Programming, Interior Point Method etc. to provide global optimal solution in bounded time.

These meta-heuristic optimization techniques find solution based on previous knowledge and number of potential solutions are tried iteratively to improve the objective function. These method use a "fitness function" to select probable solutions. These methods are probabilistic and are suitable for discrete, non-linear and non-convex functions. These methods provide global solution in the search space. However, these algorithms need problem specific information and convergence is not guaranteed. Also, different searches may field different solutions to same problem. The efficacy of the algorithm is measured in terms of memory requirement, run time and number of elementary computer operations required to solve the problem in worst case scenario.

It has been observed from the literature that heuristic and meta- heuristic techniques are widely used for solving the problem of congestion management. Among these meta-heuristic optimization techniques, PSO has been most widely used in recent years due to its various advantages as compared to other optimization techniques. It is a swarm intelligence based algorithm which uses particles through solution space having better fitness value [122]. This technique is simple as it does not involve any differential equation and also it does not require biological information. It requires less iteration to obtain optimal solution and does not get trapped in local optima. The algorithm has been discussed in detail in chapter 4 where it has been utilized to minimize cost of generation rescheduling. In this chapter, a new optimization technique based on PSO has been proposed to solve the problem of congestion management. The same problem of generation rescheduling which was considered in chapter 4 has been considered here to analyse the performance of the proposed optimization technique.

7.2 Particle swarm optimization with improved time-varying acceleration coefficients (PSO-ITVAC)

PSO is an efficient and promising optimization technique used for non-convex optimization problems which was first proposed by Kennedy and Eberhart in 1995 [122]. It uses the information of particle best position and global best position to update the position and velocities of particle as given by equations (7.1) and (7.2) respectively. The particles continue searching for solution until a convergence criteria or maximum iterations is achieved.

$$V_n^{t+1} = wV_n^t + c_1.r_1.(P_n^t - X_n^t) + c_2.r_2.(G_b^t - X_n^t) \tag{7.1}$$

$$X_n^{t+1} = X_n^t + V_n^{t+1} \tag{7.2}$$

where, w is a positive value called inertia weight

r_1 and r_2 are random values between 0 and 1

c_1 and c_2 are called acceleration coefficients and $(c_1 + c_2) \geq 4.0$.

The inertia weight w plays a significant role in algorithm convergence since it controls the impact of previous history of velocities of particles on current velocities and hence influences the local and global exploration capabilities of particles. The acceleration coefficients indicate the weighting terms which pull each particle towards personal and global best position.

PSO has found its implementation in optimization of various power system problems. It is evident from literature review that PSO has been successfully utilized to solve congestion management problem [123]-[125] using generation rescheduling based on sensitivity of generator to the power flow on congested line. In [53] too, optimal numbers of generators for congestion management were selected based on generator sensitivities to the flow of power on congested line. PSO was used to minimize the amount of active power rescheduling cost of participating generators. Rescheduling cost was calculated for different values of inertia weight. However it did not consider the effect of constriction factor and time-varying acceleration coefficients on PSO performance for calculating the rescheduling cost of participating generators. The effect of acceleration constants on PSO performance in minimizing the active power rescheduling cost of participating generators for congestion management was considered in [121]. It had developed a new algorithm for PSO in which the values of acceleration coefficients vary linearly with iteration count. However, the values of acceleration coefficients vary such that their sum was less than value of φ taken for the calculation of constriction factor k [121].

Although PSO is an efficient optimization approach as compared to other optimization methods in solving the non-convex optimization problems, its searching performance should be analyzed through its statistical results as the selection of its parameters play a vital role for its efficient operation [153]-[154].

The acceleration coefficients adjustment changes the amount of tension in the algorithm, therefore its value should be judiciously chosen as its small values will allow particles to deviate far from the target regions while its high value will cause abrupt movement of particles towards or past target regions. Generally the values of these coefficients were fixed as $c_1 = c_2 = 2.0$. But previous analysis has shown that an optimum solution can be achieved at other values of these coefficients rather than fixing them at 2.0 [153]. In [121], the authors had tried to select the values of acceleration coefficients such that it varies linearly with iteration count. But it does not consider other possibilities for optimal value of acceleration coefficients in each iteration count so that PSO converge to a more optimal and better solution. In this chapter, a modified PSO algorithm is proposed such that each particle updates its velocity with improved time-varying acceleration coefficients given by equation (7.3) and provides more optimal solution.

$$V_n^{t+1} = k \{wV_n^t + ((c_{1f} - c_{1i})\frac{t}{t_{max}} + \frac{(c_{1f} + c_{1i})^2}{3}).r_1.(P_n^t - X_n^t)$$

$$+ ((c_{2f} - c_{2i})\frac{t}{t_{max}} + \frac{(c_{2f} - c_{2i})^2}{3}).r_2.(G_b^t - X_n^t)\} \qquad (7.3)$$

where,

$$w = w_{max} - (w_{max} - w_{min})\frac{t}{T_{max}}$$

k is the constriction factor given by

$$k = \frac{2}{|2 - \varphi - \sqrt{\varphi^2 - 4\varphi}|} \; ; \qquad \varphi = (c_1 + c_2) \geq 4.0 \qquad (7.4)$$

$$c_1 = ((c_{1f} - c_{1i})\frac{t}{t_{max}} + \frac{(c_{1f} + c_{1i})^2}{3})$$

$$c_2 = ((c_{2f} - c_{2i})\frac{t}{t_{max}} + \frac{(c_{2f} - c_{2i})^2}{3})$$

Using the above modified velocity equation, the congestion management problem is optimized by the proposed technique i.e. PSO with time varying acceleration coefficients. The congestion management problem formulated in chapter-4 is considered here to analyze the performance of the proposed algorithm and is again written here.

$$\text{Minimize} \sum_{i}^{n_g} RC_i(\Delta P_{gi}) \cdot \Delta P_{gi} \tag{7.5}$$

subject to following constraints:

1. Power balance equality constraint

$$\sum_{i=1}^{n_g} \Delta P_{gi} = 0 \tag{7.6}$$

2. Operating limit inequality constraint

$$\Delta P_{gi}^{min} \leq \Delta P_{gi} \leq \Delta P_{gi}^{max} \ ; \quad i = 1, 2, \dots n_g \tag{7.7}$$

where,

$$\Delta P_{gi}^{min} = P_{gi} - P_{gi}^{min}$$

$$\Delta P_{gi}^{max} = P_{gi}^{max} - P_{gi}$$

3. Line flow inequality constraint

$$\sum_{i=1}^{N}(GS_{gi}^{pq} \cdot \Delta P_{gi}) + F_L \leq F_l^{max} \ ; \quad L = 1, 2, \dots n_L \tag{7.8}$$

where, RC_i is the rescheduling cost of generator at bus-i,

ΔP_{gi} is the active power adjustment of generator at bus-i,

ΔP_{gi}^{min} and ΔP_{gi}^{max} are respectively the minimum and maximum limit of active power adjustments of generator at bus-i,

GS_{gi}^{pq} is the generator sensitivity of generator at bus-i,

F_l^0 represents the power flow on line-l considering all the contracts,

F_l^{max} is the line flow limit of line-l connected between bus-p and bus-q,

n_t represents the total number of lines in the system.

7.3 Congestion management algorithm using PSO-ITVAC

The PSO-ITVAC optimization algorithm to find optimal solution of the objective function given by equation (7.5) with binding constraints given by equations (7.6) to (7.8) is explained as follows:

Step 1: Generate and initialize the particles with position and velocity. Every particle will have z-dimensions, z being the number of generators participating in congestion management, and the value of z variables denotes the amount of power rescheduling required by generators in order to relieve congestion.

Step 2: The binding equality constraint given by equation (7.6) and inequality constraints given by equation (7.7) and (7.8) are tested individually based on system states represented by an individual particle. If the particle does not satisfy any of the constraint, then it is regenerated.

Step 3: The optimal objective fitness values for every particle are calculated to determine the position best and global best values.

Step 4: The particles' position and velocities are updated using equation (7.2) and (7.3) respectively.

Step 5: If the pre-specified stop criterion or maximum number of iterations specified are reached, the optimization program is stopped, otherwise go to step 2.

7.4 Results and discussions

The performance of the proposed algorithm for congestion management is tested on IEEE 30-bus system, IEEE 118-bus system and a 33-bus Indian network. The results obtained for IEEE 30-bus and IEEE 118-bus systems using the proposed algorithm are compared with those reported in [121] whereas the results obtained for 33-bus Indian network are compared with the results obtained using PSO-TVAC [121]. Instead of rescheduling bids as taken in chapter-4, the cost characteristics of generators taken in [121] are considered here in order to compare the results. The simulation studies of the proposed algorithm are carried out using MATLAB. The various parameters taken for PSO is given in Table 7.1 [121].

Table 7.1 PSO-ITVAC parameters

Parameters	w_{min}	w_{max}	Φ	c_{1i}	c_{1f}	c_{2i}	c_{2f}
Values	0.4	0.9	4.1	2.5	0.5	0.5	2.5

7.4.1 Results for IEEE 30-bus system

The power flow solution gives that the congestion occurs in line connected between bus-1 and bus-2 as the power flow in it is 170.1 MW which is above its thermal limit of 130 MW [121].

The generator sensitivity (GS) values calculated for active power flow on congested line is found to be same as in [121] and is shown in Table 7.2 which reveals that all the generators have high values of GS. Therefore all generators will take part in congestion management and hence will reschedule their generation. A graphical representation of GS values calculated for IEEE 30-bus system is shown in Figure 7.1. The GS values calculated for generators of IEEE 30-bus system are utilized for calculating the amount of power re-dispatch and hence rescheduling cost using PSO-ITVAC with maximum iterations set as 500 and particle size is taken as 70. The results thus is shown in Table 7.3 which depicts that both the active power rescheduling and the total rescheduling cost obtained using PSO-ITVAC is less as compared to PSO-TVAC.

Table 7.2 Generator sensitivity values of IEEE 30-bus system for congested line connected between bus-1 and bus-2

Generator No.	1	3	5	8	11	13
GS values	0	-0.8908	-0.8527	-0.7394	-0.7258	-0.6869

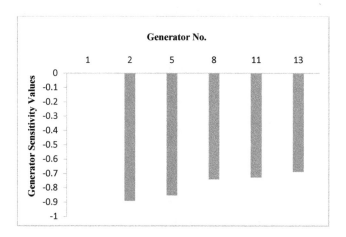

Figure 7.1 GS values for IEEE 30-bus system

Table 7.3 Results comparison for IEEE 30-bus system

Active Power Rescheduling (MW)	Proposed method	Results given in [121]
ΔP_1	-46.7	-49.3
ΔP_2	1.6	17.5
ΔP_5	9.8	14
ΔP_8	10.2	9.9
ΔP_{11}	20.9	6.8
ΔP_{13}	4.3	3
Total ΔP	93.4	100.5
Cost ($/hr)	240.9	247.5

7.4.2 Results IEEE 118-bus system

The load flow solution for IEEE 118-bus system gives that line connected between bus-89 and bus-90 is congested as the power flow in it is 260 MW which is above its power flow limit of 200 MW [121].

The GS values of all generators calculated for congested line comes out to be same as [121] and is shown in Table 7.4 which reveals that for this congested line, generators connected at bus-85, bus-87, bus-90, bus-89 and bus-91 have high magnitude of GS as compared to other generators as illustrated from Figure 7.2 too.

Table 7.4 Generator sensitivity values of IEEE 118-bus system for congested line connected between bus-89 and bus-90

Gen. No.	GS Values (10^{-3})	Gen. No.	GS Values (10^{-3})	Gen. No.	GS Values (10^{-3})
1	0	42	-0.0375	80	-0.9250
4	-0.0005	46	-0.0242	85	**50.068**
6	-0.0001	49	-0.0460	87	**50.654**
8	-0.0014	54	-0.0838	89	**74.455**
10	-0.0014	55	-0.0871	90	**-701.153**
12	0.0004	56	-0.0854	91	**-427.901**
15	0.0021	59	-0.1100	92	-28.411
18	0.0051	61	-0.1160	99	-9.391
19	0.0046	62	-0.1130	100	-12.915
24	0.1350	65	-0.1350	103	-12.737
25	0.0484	66	-0.0983	104	-12.854
26	0.0337	69	0.2120	105	-12.772
27	0.0451	70	0.3690	107	-12.202
31	0.0339	72	0.2326	110	-12.274
32	0.0477	73	0.3400	111	-12.070
34	-0.0323	74	0.5410	112	-11.747
36	-0.0329	76	0.8650	113	0.0110
40	-0.0343	77	0.0012	116	-0.1750

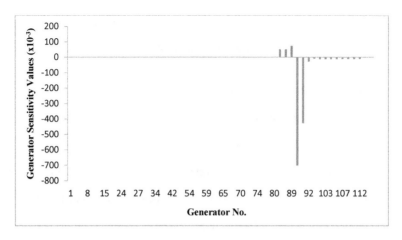

Figure 7.2 GS values for IEEE 118-bus system

Therefore a significant change in these generators output will affect the power flow on congested line and hence these generators will participate in congestion management along with slack bus generator which account for the system losses. The amount of active power rescheduling of selected generators and total cost of rescheduling using proposed algorithm with maximum iteration count set to 1000 is shown in Table 7.5. The particle size is taken as 70.

Table 7.5 Results comparison for IEEE 118-bus system

Active Power Rescheduling (MW)	Proposed method	Results given in [121]
ΔP_1	-3.2	-4.4
ΔP_{85}	-3.9	-10.3
ΔP_{87}	-4.3	-22
ΔP_{89}	-68.1	-58.5
ΔP_{90}	60.9	69.4
ΔP_{91}	18.6	24.7
Total ΔP	159.1	189.3
Cost ($/hr)	896.0	970.7

The generation rescheduling cost using PSO-ITVAC comes out to be 896.0 $/hr which is less as compared to 970.7 $/hr obtained with PSO-TVAC.

7.4.3 33-bus Indian network

A 33-bus Indian network is also taken to test the feasibility of proposed algorithm. It consists of 9 generator buses, 24 load buses and 46 transmission lines. The power flow solution gives that line connected between bus-8 and bus-23 is congested as the power flow through it is 47.2 MW above its power flow limit of 400 MW.

Table 7.6 shows the GS values of all generators calculated for the congested line and it depicts that generators connected at buse-5, bus-12, bus-23 and bus-24 have high magnitude of GS. Therefore only these four generators along with slack bus generator will participate to relieve congestion as illustrated in Figure 7.3.

Table 7.6 Generator sensitivity values of 33-bus Indian network for congested line connected between bus-8 and bus-23

S. No.	Gen. No.	GS Values
1	1	0
2	2	-0.0001
3	5	**-0.0642**
4	12	**0.0583**
5	17	0.0027
6	23	**-0.1268**
7	24	**0.078**
8	32	-0.0018
9	33	-0.0001

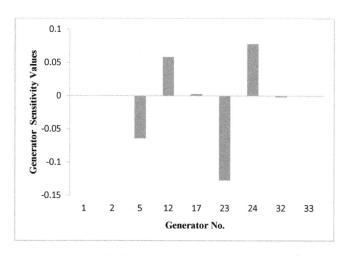

Figure 7.3 GS values for 33-bus Indian network

The amount of active power required for each participating generator to reschedule is calculated using the proposed algorithm and is given in Table 7.7. The maximum iteration count is set to 1000 and the particle size is taken as 70. The cost coefficients of participating generators are same as of IEEE 30-bus system generators.

Table 7.7 Results comparison for 33-bus Indian network

Active power rescheduling (MW)	Proposed method	Method proposed in [121]
ΔP_1	-17.6	-34.3
ΔP_5	28.3	23.1
ΔP_{12}	-9.8	-5.2
ΔP_{23}	15.2	25.4
ΔP_{24}	-16.1	-9.0
Total ΔP	86.9	97.0
Cost ($/hr)	221.6	246.7

The total rescheduling cost calculated using PSO-ITVAC is compared with those obtained using PSO-TVAC and it is found that the total PSO-ITVAC gives less generation rescheduling cost (221.6 $/hr) as compared to that with PSO-TVAC (246.7 $/hr).

A comparison of the total rescheduling cost using PSO-ITVAC and PSO-TVAC for different systems taken for analysis is illustrated in Figure 7.4.

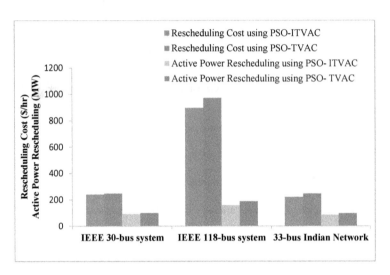

Figure 7.4 Rescheduling cost and active power rescheduling for different systems

7.4.4 Performance characteristics of PSO-ITVAC

The selection of PSO parameters (i.e. population size, inertia weight, acceleration constants etc.) plays a quite significant role in PSO performance as it has great influence on PSO convergence characteristic. Therefore these parameters should be judiciously and cautiously selected in order to achieve better performance of PSO. In order to compare the PSO-ITVAC performance with PSO-TVAC, the same values of these parameters were taken as in [121] and are given in Table 7.1. As PSO is a stochastic optimization technique, a random population of particles is generated in each new simulation thereby giving almost different results in each simulation. Therefore 50 trial simulations were carried out and the maximum, minimum and average value of active power rescheduling and total rescheduling cost were noted down. The values thus obtained using PSO-ITVAC are compared with PSO-TVAC for IEEE 30-bus and IEEE 118-bus system as shown in Table 7.8 and Table 7.9 respectively.

Table 7.8 Statistical results for IEEE 30-bus system

MW Rescheduling	Proposed method			Results reported in [121]		
	Max	Min	Average	Max	Min	Average
ΔP_1	-51.5	-41.7	-46.7	-51.5	-47.3	-49.3
ΔP_2	23.7	0.2	1.6	22.0	25.1	17.5
ΔP_5	18.3	11.0	9.8	14.7	16.0	14.0
ΔP_8	3.2	5.2	10.2	8.8	7.6	9.9
ΔP_{11}	1.4	22.5	20.9	6.2	0.6	6.8
ΔP_{13}	4.8	2.8	4.3	1.0	0.0	3.0
Total ΔP	103.1	83.4	93.4	103.8	96.7	100.5
Rescheduling Cost ($/hr)	233.7	214.8	240.9	254.9	237.9	247.5

Table 7.9 Statistical results for IEEE 118-bus system

MW Rescheduling	Proposed method			Results reported in [121]		
	Max	Min	Average	Max	Min	Average
ΔP_1	-6.8	-1.1	-3.2	-5.9	-0.8	-4.4
ΔP_{85}	-7.4	-2.3	-3.9	-6.2	-12.1	-10.3
ΔP_{87}	-18.2	-4.6	-4.3	-6.5	-13.9	-22.0
ΔP_{89}	-63.3	-65.9	-68.1	-96.2	-52.3	-58.5
ΔP_{90}	59.6	61.5	60.9	80.1	81.6	69.4
ΔP_{91}	36.1	12.4	12.4	30.5	3.3	24.7
Total ΔP	191.4	147.8	159.1	225.5	163.8	189.3
Rescheduling Cost ($/hr)	1093.2	765.9	896.7	1229.6	829.5	970.7

The maximum, minimum and average results of active power rescheduling and rescheduling cost obtained for 50 trial simulations for 33-bus Indian network using PSO-ITVAC and PSO-TVAC is also obtained as shown in Table 7.10.

Table 7.10 Statistical results for 33-bus Indian network

MW Rescheduling	PSO-ITVAC			PSO-TVAC [121]		
	Max	Min	Average	Max	Min	Average
ΔP_1	-22.0	-14.2	-17.6	-46.3	-33.8	-34.3
ΔP_5	29.2	18.7	28.3	29.4	16.0	23.1
ΔP_{12}	-27.3	-19.8	-9.8	-5.3	-8.5	-5.2
ΔP_{23}	24.1	15.9	15.2	29.8	27.4	25.4
ΔP_{24}	-4.0	-0.5	-16.1	-7.5	-1.1	-9.0
Total ΔP	106.7	69.1	86.9	118.3	86.7	97.0
Rescheduling Cost ($/hr)	281.6	167.7	221.6	302.6	215.8	246.7

The convergence characteristics of PSO-ITVAC for IEEE 30-bus system, IEEE 118 bus system and 33-bus Indian network are shown in Figure 7.5, Figure 7.6 and Figure 7.7 respectively which illustrate that the total rescheduling cost decreases with iteration count and finally settle to a minimum value giving the optimum solution. Figure 7.5 and Figure 7.7 illustrates that for a small network, PSO-ITVAC is more efficient than PSO-TVAC as the total rescheduling cost decreases with iteration count and finally settles to a minimum value giving the optimum solution. Figure 7.6 illustrates the effectiveness of the proposed algorithm and reveals that for a large network too, the proposed algorithm is more efficient than PSO-TVAC in minimizing the total rescheduling cost of the participating generators for congestion management. It is also evident from the figures that the PSO-ITVAC converges more rapidly and gives less rescheduling cost as compared to PSO-TVAC regardless of the size of network. Thus it can be inferred that PSO-ITVAC is more efficient in minimizing the rescheduling cost as compared to PSO-TVAC.

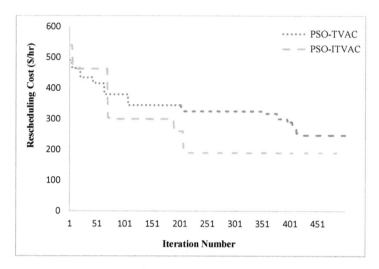

Figure 7.5 Convergence characteristics of PSO-ITVAC and PSO-TVAC for IEEE 30-bus system

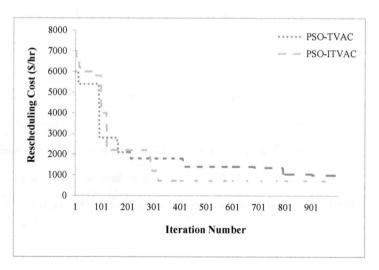

Figure 7.6 Convergence characteristics of PSO-ITVAC and PSO-TVAC for IEEE 118-bus system

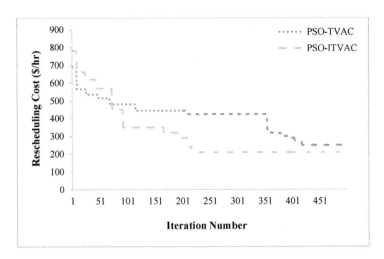

Figure 7.7 Convergence characteristics of PSO-ITVAC and PSO-TVAC for 33-bus Indian network

7.5 Conclusion

An efficient optimization algorithm based on PSO has been proposed to solve the congestion management problem. The performance of the proposed optimization algorithm, PSO-ITVAC, has been analysed considering the generation rescheduling method of congestion management. It has been tested on IEEE 30-bus system, IEEE 118 bus system and 33-bus Indian network.

It has been observed that the PSO-ITVAC manages congestion more efficiently as compared to PSO-TVAC method of optimization. Also, it has been found that active power rescheduling cost using PSO-ITVAC is more efficiently minimized as compared to PSO-TVAC for small as well as large networks. Furthermore, PSO-ITVAC converges to optimal solution more rapidly than PSO-TVAC.

CHAPTER-8

CONCLUSION AND FUTURE SCOPE

8.1 Introduction

Different market structures have been evolved due to introduction of competition in electricity market which results in low electricity cost. However, the shift of market structure from monopolistic to deregulated has raised certain technical issues which could hamper the operation of power system. Also, it could take away the benefits which the deregulation of electricity industry promises to the society. Secure operation of power systems has been a challenging task in bundled as well as unbundled structure of electricity supply industry. To achieve secure operation of power system network is somewhat complex in deregulated market as compared to bundled power system.

As congestion is central to the issue of transmission tariffs, losses as well as market economics, therefore congestion management remains the central issue to power transfer management in competitive environment of electricity market. Market participants do not bother about the reliability and security of the system. They participate with the only aim of maximizing their profit. Without robust congestion management strategy, their actions can put the transmission system operation in stake. Therefore, interaction of congestion management with energy market economics should be carefully accomplished such that market inefficiency cannot take away the benefits promised by deregulation to the society. This thesis has addressed the issue of congestion management of transmission network under competitive environment of electricity supply industry.

8.2 Contributions of the research

The main contributions of the work can be summarized as follows:

1. Three different congestion management methodologies have been discussed. Each method discussed in the thesis manages congestion efficiently. However, none of these methodologies could be claimed superior to each other as their adoption for congestion management depends upon the market scenario and the system requirement.

2. A PoolCo model of electricity market structure has been adopted for management of power flow in transmission network.

3. A generator sensitivity based congestion management mechanism using generation rescheduling has been formulated which also considers generator's rescheduling bids for optimal rescheduling of power output of generators. By introducing rescheduling bids along with generator sensitivity gives more optimal rescheduling cost. The effect of change of rescheduling bids of generator has also been investigated which shows that varying the rescheduling bid of a generator accordingly changes its power generation for congestion management thereby providing the minimal rescheduling cost.

4. The impact of distributed generation in managing congestion has been investigated. A zonal congestion management approach using distributed generation has been adopted. The identification of different congestion zones is based on LMP difference between buses connecting a line. The most sensitive congestion zone is one which groups the buses connecting lines of high and non-uniform LMP difference across them. The congestion is managed by optimally allocating the DG in most sensitive congestion zone. The optimal allocation of DG is also done based on LMP difference which gives the minimum generation

cost as compared to other potential locations of DG allocation. The methodology is found to be efficient for both small as well as large power systems.

5. The impact of FACTS device in managing congestion of transmission lines has also been analyzed wherein TCSC due to its low installation cost has been considered for congestion mitigation. A static model of TCSC is considered in problem formulation. The optimal location of TCSC is determined so as to reduce the effective cost of its installation as well as to reduce the cost of generation so that congestion could be managed efficiently. The determination of optimal location of TCSC is done based on line flow sensitivity factor calculated for each line. TCSC is placed in most sensitive line having the most negative value of line flow sensitivity factor that gives minimum installation cost of TCSC as well as minimum generation cost as compared to other potential locations for TCSC placement.

6. An efficient optimization algorithm based on PSO has been developed to solve the congestion management problem. The performance of the developed optimization algorithm, PSO-ITVAC, has been analyzed considering the generation rescheduling method of congestion management. The developed algorithm, PSO-ITVAC, manages congestion more efficiently as compared to PSO-TVAC method of optimization. Also, the cost of active power rescheduling using PSO-ITVAC is more efficiently minimized as compared to PSO-TVAC for small as well as large networks. Furthermore, PSO-ITVAC converges to rapidly to a more optimal solution than PSO-TVAC.

7. In developing countries, almost all generating units work near to its rating and have low reserve margin. Therefore, generation rescheduling method of congestion management is not encouraged to be applied if all demands have to be

served. Hence, other two methods of managing congestion are suitable in those situations.

8.3 Scope for future research

The present work can be extended to include the following aspects:

1. In the present work, minimization of generation cost is estimated by rescheduling real power outputs of generators. This study can be extended to include reactive power outputs.
2. The developed algorithm for congestion management makes use of only bids received from generating companies. It can be extended to include demand side bids.
3. The allocation of distributed generation for congestion management may further be investigated for selection of optimal size of distributed generation.
4. The impact of other FACTS devices such as SSSC, STATCOM, UPFC etc. in managing congestion can also be carried out.
5. Management of congestion can further be investigated in a market structure having combination of both PoolCo and bilateral or multilateral trade agreements.
6. A comparative analysis between different optimization algorithms for solving congestion management problem can be further carried out.

CPSIA information can be obtained
at www.ICGtesting.com
Printed in the USA
BVHW030845091222
653835BV00015B/312